T0305135

Achieving Business Success with GIS

Achieving Business Success with GIS

BRUCE DOUGLAS
Director, Corporate GIS Consultants, Australia

John Wiley & Sons, Ltd

Other Wiley Editorial Offices

John Wiley & Sons Inc., 111 River Street, Hoboken, NJ 07030, USA

Jossey-Bass, 989 Market Street, San Francisco, CA 94103-1741, USA

Wiley-VCH Verlag GmbH, Boschstr. 12, D-69469 Weinheim, Germany

John Wiley & Sons Australia Ltd, 42 McDougall Street, Milton, Queensland 4064, Australia

John Wiley & Sons (Asia) Pte Ltd, 2 Clementi Loop #02-01, Jin Xing Distripark, Singapore 129809

John Wiley & Sons Canada Ltd, 6045 Freemont Blvd, Mississauga, ONT, L5R 4J3, Canada

Wiley also publishes its books in a variety of electronic formats. Some content that appears in print may
not be available in electronic books.

Library of Congress Cataloging-in-Publication Data

Douglas, Bruce, 1952–
   Achieving business success with GIS / Bruce Douglas.
      p. cm.
   Includes index.
   ISBN 978-0-470-72724-9
   1. Geographic information systems—Management.   2. Information technology—Management.
I. Title.
   G70.212.D68 2008
   658.05—dc22

                                                                        2007033358

Typeset in 10.5/12.5pt Times by Aptara Inc., New Delhi, India.
Printed and bound in Great Britain by Antony Rowe Ltd, Chippenham, Wiltshire
This book is printed on acid-free paper responsibly manufactured from sustainable forestry in which at least two
trees are planted for each one used for paper production.

Contents

Preface

Geographic Information Systems (GIS) are a mainstream technology with a vital and growing use across all industries, including natural resources, oil and gas, military, environment, education, insurance, transport and logistics, land administration, utilities and many more. Geographic (or geospatial / spatial) Information Systems[1] are typically focused on storing data which has a geographic location, undertaking analysis of that data, integrating that data with other data types and presenting that data for decision support, usually in the form of a map.

GIS has been used since the late 1970s by an increasing number of organisations. However, in a large number of cases it remains under-utilised and often has not provided the business benefits originally envisaged. Of course, most software vendors would say that any such problems could be solved by buying their product – but this is not correct of course. To make 'GIS work' requires more than just a simplistic software approach – it requires understanding the business environment of the host organisation and how this technology can be used to address real business issues.

This book explores the *business* environment of making GIS successful. It does not discuss specific technology per se, but provides a business-focused rationale for using spatial technologies to address real business problems.

If the reader is seeking a book about the specific functionality of GIS software (or genres of software), then this is not the book for you. There are a number of excellent books available which examine the technology of GIS, delve into the different aspects of the software, look at polygon management, discuss layers, focus on Boolean logic, explain how linear topology and dynamic segmentation work, describe web feature services, discuss database concepts and so on. This book does not cover any of these technical issues, except in broad terms, and only then on the basis of how these issues might assist the GIS to successfully deliver business outcomes.

This book discusses business issues, that is, how to develop a GIS strategy, how requirements should be defined, how to select and implement the most *appropriate* GIS, the issues which need to be considered to ensure success when using GIS and issues relevant to the spatial industry. Above all, this book seeks to provide the mechanism so that spatial technologies support and enhance the delivery of business benefits to the organisation implementing the technology.

As management consultants in the spatial information industry, Corporate GIS Consultants are regularly involved in working with organisations (mainly government agencies and utilities) to determine whether the spatial information technology used in the subject organisation meets actual business requirements, and, if it does not, to determine what needs to be done so that it can meet these business needs.

While many organisations have used GIS for a number of years, and in some cases invested many millions of dollars in this technology, a high percentage have yet to gain

---

[1] GIS, Geographic, GeoSpatial and Spatial all generally mean the same thing and are used interchangeably in this book.

appreciable business benefits. This may be due to a number of issues such as the business needs not being fully defined in the first place, the impact of GIS on current / potential business processes not being fully understood, staff not being adequately trained, etc. An incorrect business focus on GIS may result in the usefulness of this technology being sub-optimal.

Why geographic or spatial? It is because a picture tells a thousand words. This is true for computer-based systems as well as paper-based systems. The human brain is excellent at processing images and drawing solutions from complex spatial patterns. Spatial systems can present complex numerical data as simple images which the human brain can more easily process to provide the basis for a better understanding of complex situations involving lots of data and, hopefully, to make better decisions.

This book is focused on helping readers to achieve business success with their GIS, and would therefore be just as appropriate to the business unit line-manager who may be responsible for the GIS as well as the student trying to understand how to make GIS successful. I trust that this book is useful.

Bruce Douglas

Acknowledgements

The development of this book has been a lengthy and sometimes painful process, particularly when one also has a day job. I would like to thank a number of friends and business colleagues who have provided assistance, support and encouragement for me to continue with this book.

In particular, I would like to thank Carole Lowe, my business partner and co-Director of Corporate GIS Consultants, for her support and assistance with the book, Jon Fairall from South Pacific Science Press for his suggestions on the content and structure of the book and Thierry Gregorius from Shell in The Netherlands for his detailed review of the book. In addition, many thanks to the publishers and their reviewers who thought that this book had some merit and should be published.

Bruce Douglas
June 2007

1 Introduction

GIS technology has been used in a variety of government and semi-government organisations since the late 1970s. However it has only been in the last two decades that GIS has become a technology which has permeated almost all facets of government, utilities and commercial organisations which have an interest in matters relating to:

- land (e.g. ownership, leasing/licensing, etc.);
- services on, under or over the land (e.g. utility services – pipes, drains, power networks, telecommunications, etc.);
- commercial activities on or about the land (e.g. mining, agricultural, land development, property, banking, real estate, transportation, environmental, planning, etc.); and
- other matters which relate to land and property (e.g. policing, emergency services, etc).

As such, GIS has become a pervasive technology in a number of organisations. Along the way GIS has also become mainstream and now sits alongside Finance Systems, Human Resource Systems, Asset Management Systems and Customer Management Systems in the Information Technology (IT) environments of government departments, utilities and private companies.

GIS now typically runs on Microsoft® Windows™ platforms on corporate networks and most desktops – a significant change from 5 or 10 years ago when GIS used to be very expensive technology running on isolated Unix-based workstations/networks used only by highly trained gurus. The demise of the use of Unix for GIS is shown in the following chart.[1] This can also be directly correlated with the use of the Windows™ operating system as the corporate desktop and the migration of most GIS software to this platform.

Since the 1970s, GIS has been used to computerise basic land parcel information, but has since grown over the last couple of decades to include a plethora of applications, such as using GIS for:

- managing assets and services;
- managing land-related information;
- managing customers for service delivery (whether it is for delivery of water services or postal items);
- integrating information from other (often disparate) systems;
- inclusion into Police and Emergency Services computer-aided dispatch systems;
- modelling transportation systems; and
- monitoring and management of environmental and forestry issues.

---

[1] Source: GIS/Spatial Best Practice Surveys, Australia and New Zealand, Corporate GIS Consultants 2000–2006.

Achieving Business Success with GIS   Bruce Douglas
© 2008 John Wiley & Sons, Ltd   ISBN: 978-0-470-72724-9

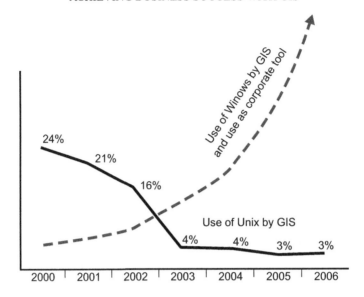

As a technology, GIS is principally focused on:

- capturing/converting/storing spatial or geographic data (sometimes also called geospatial data), i.e. data which has a location on, over or under the surface of the earth;
- undertaking analysis of that data;
- integrating that data with other data types; and
- presenting that data for decision support, generally in the form of a map, perhaps with associated tables or charts.

Therefore a GIS can be defined as a computer system that stores, manages, displays, analyses and reports on information which has a 'where' component, i.e. a location, because a number of decisions undertaken by staff in land-related organisations often refer to 'what', 'where', 'how much', 'what does this relate to' and 'how does this relate to my other data'. The inclusion of the 'where' component into the decision-making process can be a powerful tool for providing a better understanding of the issues at hand and the implications of specific decision paths.

GIS environments most often consist of two major components: the spatial or graphic data which represents real-world entities; and the aspatial or textual data which describes the attributes of those real-world entities. This data is then accessed by software which can relate each item of both the graphic and textual data to each other to provide information to the user.

Because spatial data is just that – spatial – a key feature of a GIS environment is that each spatial entity intrinsically knows the relationship of itself to other spatial entities. For example, the relationship of a street to a bus stop on that street can be derived from the GIS without the need to have a specific relationship defined in the database that relates the particular bus stop to the specific street.

Case example

Spatial systems do not need to have relationships built to link data as do traditional IT systems which link data by 'common keys' in a textual database.

On a recent very large project (70 developers) to implement a national GIS, we found that some of the IT developers were designing and developing a method to build relationships between spatial data in a traditional Relational Database without understanding that they had a very sophisticated GIS system which did just that. Needless to say, I stopped them from doing this and arranged for them to have the requisite training so that they understood how spatial systems worked.

A GIS typically contains a considerable amount of spatial data, often referenced to a 'map base' containing data on topography, property, roads, railways, parks etc. to provide a map context to the data being used. The map base is usually a mathematically defined representation of the surface of the earth, typically represented as a map projection stored as Cartesian coordinates such as easting and northing or Latitude and Longitude coordinates.

The most obvious advantage of representing 'where' data in a GIS can be derived from the visual representation of data presented in map form. When this is also combined with other information about the data item, such as what the entity is, how much it costs, what it relates to, etc., more powerful relationships can be determined. For example, the location or spatial distribution of socio-economic data (e.g. income) can be a very useful tool in planning for the provision of government services, etc.

As such, the inherent spatial relationships within the map data can provide powerful benefits that might not be initially obvious, such as:

- finding the relationship of different data themes – e.g. '*show me the spatial distribution of car ownership compared with bus route locations for these suburbs*';
- finding items at a given location – e.g. '*show me all the planning constraints on a property at this location*';
- finding locations where certain conditions are met – e.g. '*show me where all the contamination is within 100 metres of this watercourse, particularly that contamination that has not been remediated within the last 5 years*';
- identifying trends and spatial patterns – e.g. '*show me all water pipes that have failed more than twice in the last 2 years and correlate this information with telemetry data pertaining to pipe pressure*';
- scenario modelling – typically used to assess or distinguish between a set of proposed options, for example assessing the merit of installing new infrastructure as opposed to upgrading existing infrastructure.

So GIS can be a powerful tool to represent and analyse spatial data. But in order to do this, the GIS must have a lot of data captured, and that data must be relevant to the queries being run (i.e. a query involving planning constraints must use the latest planning data) and must be as correct/accurate as possible.

Thus, although the business benefits that can be obtained from using GIS are significant, these benefits will only happen if the GIS has good data. GIS is a data-centric application which can provide substantial business benefits if, *and only if*, the data is available and it has a high degree of correctness and is suitable for the purpose to which it is going to be used.

In addition, when this good GIS data is combined with data from other systems (e.g. customer data, sales data, financial data, etc.), substantially more business benefits can be derived. All of this information, properly presented, can present a unique view on the data world, leading to this data becoming information which is then able to be used to derive knowledge to augment a number of business outcomes, and in some cases to provide business intelligence.

However, in the process of implementing/reviewing GIS, organisations often focus on the GIS software on the assumption that the software is the key to 'solving GIS' and that if the right software is purchased then all the rest will be easy. This is an incorrect assumption of course, but nevertheless one that I am frequently asked – i.e. 'which is the best system'. The best system, of course, depends on what the system is going to be used for. Of the dozen or so GISs available on the market, probably over half of these would suit the functional requirements for most organisations.

Fact

There is no best GIS – just as there is no best motor vehicle. It all depends on the requirements of the user, the degree of sophistication required and the size of the budget. And just as the best motor vehicle for one user may be a heavy-duty 4WD (such as a Toyota Land Cruiser), the best motor vehicle for another user may be an up-market sports car (e.g. a BMW Z4) – each will have substantially different prices, substantially different functionality and substantially different business fit (quite apart from substantially different sex appeal) to meet the needs of the prospective owner (and his/her family).

Thus, there is no such thing as a best GIS. It all depends on the needs of the organisation, the use to which the system will be put, the data that is available to use with the system and the integration of the GIS with the other corporate systems. We are often asked whether the spatial information technology used by a client organisation is appropriate to their needs, and, if not, what needs to be done in order to make the technology more focused to the needs of the business. Often this is because organisations do not understand their current (and potential) business processes that may use GIS or the full capability of spatial information technology and hence do not understand the benefits that spatial information technology may have on the business.

GIS, like any other technology, should be based on meeting the business needs of the organisation, and all organisations, including public agencies, have business needs. Mostly these business needs will be expressed as corporate objectives or operational goals. The Annual Report of an organisation is always a useful place to start to review the Corporate Charter, i.e. the key goals, objectives, performance indicators, result areas and service commitments needed to fulfil the expectations of shareholders – particularly if the government is the major shareholder (e.g. a government agency or government-owned business).

Good business planning often results in organisations defining Key Performance Indicators (KPIs) which can be used to measure the degree of compliance with these goals and objectives – i.e. 'goalposts for the business'. How an organisation 'measures up' with respect to its KPIs is often a major factor in determining success. Meeting KPIs is also useful to gain additional budget or to obtain approval for new initiatives. As such, KPIs are often *critical to the longevity* of organisations or of staff in those organisations. And if an organisation does not meet its goals and objectives (or 'measures up' poorly against its KPIs), the question 'why not?' often follows – sometimes with disastrous outcomes.

Therefore all technology, particularly GIS, must be focused on meeting the needs of the business if it is to be successful. Why 'particularly' GIS? Because GIS needs a lot of good data for it to be really useful to the business and most of this data is usually spread across the organisation, in lots of different organisational silos. Unlike traditional IT systems, such as Financial Systems or Human Resource Systems which are often single-focus, GIS usually is multi-focus across an organisation and often requires organisational issues to be addressed in order for the implementation to be successful. And because GIS and related spatial technologies can be a powerful mechanism to assist organisations to meet KPIs, their use is becoming more widespread and more functionally rich.

If GIS is properly implemented it should be able to bring substantial benefits to the organisation. But this requires that the benefits be measured. This also requires that one knows the level of performance of these KPIs before the technology is installed, as well as being able to measure them after the GIS has been running for some time. That is, one needs to know where the goalposts are and when a goal has been scored.

To do this it is important to have:

- a strategy or 'road map' to ensure that the use of the spatial technology is aligned with the organisation's business objectives and that the benefits that the technology might be able to bring to the business (and in the right timeframe) are identified;
- methodologies and business practices to implement the strategy and to manage the changes it will bring; and
- appropriate mechanisms to 'measure' the organisation's use of the technology, both at any given point in time and as it progresses toward reaching the overall goal(s). Internal measures (to measure against internal expectations and/or teams) and external measures (to measure against other 'like' external businesses to ensure competitive edge is maintained) are essential.

In essence, this book therefore provides an overview of the *business basis* for implementing and using GIS.

The following chapters provide a review of the GIS/Spatial Information market predominantly in Australia and New Zealand, but also with reference to other countries in order to provide a contextual background for further discussions and consideration of key issues which will impact GIS in a typical organisation. All of these issues are then typically used as an input into the development of a GIS Strategy.

The Business focus, Data/Information focus, Organisational focus and Application/Technology focus all lead into the development of the GIS Strategy. For this strategy to be successful, it must meet the business needs of the organisation – Chapter 8 draws all the issue together into the strategy.

The Cost/Benefit Analysis (and Business Case) discussed in Chapter 9 is very important to gain approval of the overall strategy/roadmap and to gain access to the appropriate budget to implement the strategy. After further defining, refining and refocusing of the strategy and business needs, the selection of the most appropriate GIS technology can be undertaken, typically by a tender process. And at the end of this process, the system is (finally) implemented. But is this the end of the process, or just the beginning of a much larger process?

The following chapters provide a commentary on the best way to undertake the above tasks so that this process, and the resultant spatial technology, is focused on using it (the GIS) to deliver real outcomes for the business. Along the way, a number of case studies are cited to emphasise specific points.

2 The Spatial Information Industry

In any discussion about the Spatial Information (SI) industry, and indeed any discussion about GIS technology, it is important to understand a number of issues such as:

- the application of GIS to meet the needs of organisations;
- the market in which the technology operates;
- the GIS products commonly used in the marketplace;
- the 'value' of the industry;
- why organisations buy this technology;
- what problems they hope to solve with it;
- how innovative the suppliers of this technology are; and
- what software development is underway.

In addition, it is also useful to understand the initiatives undertaken by the government agencies in different jurisdictions, in that their strategies often have a substantial impact on the data availability for those regions.

In 2000, Corporate GIS Consultants developed an industry survey, the 'GIS/Spatial Best Practice Survey' specifically focused on Australia and New Zealand. The aim of this survey was to obtain and provide credible and relevant statistics about the industry so that any advice offered, or recommendations resulting from a project, could be made with full knowledge of 'best (available) practice'. This survey has been undertaken annually since then, generally during the March and April timeframe of each year.

While this survey has expanded over the last 7 years to include responses from a number of other countries, the focus remains on Australia and New Zealand. However, in developing this book further research has been undertaken to source other surveys in the USA and elsewhere and these surveys are suitably referenced where mentioned. In particular, information is cited from the Geospatial Technology Report, GITA,[1] North America, where appropriate.

The need for this survey evolved from the lack of accurate (or any) statistics about the use of GIS by major industry segments (e.g. Government, Utilities, etc.) in a number of countries. Market surveys and publications are often very difficult to extrapolate from marketplaces in other countries because:

- the software suppliers that are represented in countries such as the USA, Japan, Europe and Australia are in some cases quite different from suppliers in other countries – i.e. each country has local products which are not sold in other countries, and vice versa;[2] and

---

[1] GITA – Geospatial Information and Technology Association.

[2] For example, several major GIS suppliers in the USA and Europe are not represented in Australia and New Zealand and of the (approximate) dozen major GIS suppliers in Japan over half do not have customers (or a presence) external to Japan.

Achieving Business Success with GIS   Bruce Douglas
© 2008 John Wiley & Sons, Ltd   ISBN: 978-0-470-72724-9

• the products which are sold in the majority of other countries are likely to have substantially different 'market shares' and support arrangements to those offered in this part of the world.

However, a general observation would be that while individual suppliers and market shares might vary, the general principles and market trends found in the Australian/New Zealand surveys can be extrapolated to other countries. Therefore for the last 7 years, this GIS Survey has uniquely provided statistics for the Australian and New Zealand industry which have been used as the basis for comparison by organisations wanting to improve the performance of their GIS environment.

The objectives of the GIS Survey are to determine a level of 'Best Practice' for the use of spatial and related technologies across a broad range of implementation considerations that can impact business directions, so that industry practitioners can better manage SI technology to meet their specific business needs. In this context, 'Best Practice' can be considered as being that practice which is at the upper level of 'Common Practice'.

2.1   BACKGROUND TO THE SURVEY

Contributions to this survey are voluntary and only allowed from bona fide 'end-users', i.e. contributions from vendors and consultants (who sell systems) are not accepted. There have been 1,985 contributions from organisations using GIS for the 7 years from 2000 to 2006 – generally at a rate of between 450 and 550 each year (although at a lesser rate in earlier years). In each year there is approximately a 60% retention rate of respondents (from the previous year) and 40% new contributors. This provides a healthy mix of new and continuing contributors.

All contributors receive a free copy of a Contributor Report in return for their time and input, this free report being the only incentive provided to gain responses to the survey – there is no offer of any financial incentive to garner contributions to the survey, nor offers of inducements such as 'go into a draw to win a personal navigation system'.

Given these statistics and knowledge of the numbers of users who constitute each 'vendor community', it is estimated that there are between 3500 and 4500 GIS sites in the SI industry in Australia and New Zealand.

Since 1,985 contributors have responded over the 7 years of this survey, and allowing for multiple systems on a large number of contributor sites, over the last 7 years it is estimated that between a third to a half of the total SI community have responded to the survey and that for each year approximately a quarter of the industry have responded. As such, over the last 7 years the GIS survey has provided a longitudinal database which represents a unique 'statistical slice' of the SI industry.

The entire survey process is now undertaken on the Internet where new contributors can register to complete the survey and received a username and password which provide them access to the survey.

The percentage contribution rates across industry sectors and industry regions vary slightly from year to year, but are generally of the order as shown in the following charts, both in terms of industry sector and the regional response.

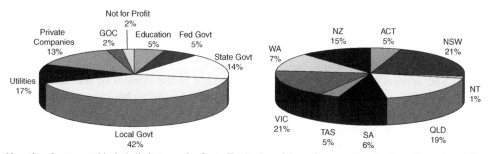

Note that the geographic jurisdictions are for States/Territories of Australia as well as for New Zealand, and that the industry segmentation is for Federal Government (Australia and New Zealand Federal Governments), State Government, Local (Municipal) Governments, Education, Utilities, Not for Profit Organisations, Private Companies and Government-Owned Corporations (GOC).

This survey asks questions on, and has consequently measured a number of indicators across, a number of domain areas over the last 7 years, as follows:

- Board Industry Trends – i.e. GIS product usage, operating system, product deployment architectures, data storage, infrastructure strategy, computer-aided design (CAD) usage and previous GIS usage.
- Budgets – i.e. overall industry value, industry growth, budgets for each industry sector, next year's budget, etc.
- Spatial Information Applications – for each industry sector.
- Training – for each industry sector.
- System Integration – for each industry sector.
- Spatial Data – for each industry sector, spatial data suppliers, issues impacting the Australian Spatial Data Directory (ASDD), data pricing policies and data liability and privacy.
- Imagery – Imagery products, processing, user perceptions and future use.
- Mobile Computing – mobile computing use and products, business benefits, success factors and mobile solution risks.

As part of the 'commencement process' of the survey each year, input is sought from a small review panel consisting of a number of 'key industry practioners' inviting comment and relevance of questions asked, etc. While occasionally there have been some new questions posed, or other minor comment, the survey has 'stood the test of time' and, judging by its level of support, remains relevant to the industry.

2.2 VALUE OF THE SI INDUSTRY

The monetary 'value' of 'the industry' is always challenging and typically depends on the determination of two questions: how is 'the industry' defined? and how is 'the value' defined? These issues are discussed below.

1. How is 'the industry' defined?
   There have been many definitions of 'the GIS industry' proposed by different vested interest groups over the last decade. Some of these groups have tried to define 'the industry' by narrow categorisations, e.g. the 'survey and mapping' sector, the

'natural resources' sector, etc. Often these categorisations have been impacted (and in some cases, impeded) by local debates about the industry based on particular 'stand-points' which, in general have not gained full acceptance by the other sectors, i.e. the disenfranchised sectors.

The approach that we undertook on this issue in the context of the survey was to turn the question around and to let those who use the technology and therefore would consider that they are in 'the industry' to respond to the survey, with the responses being subsequently analysed to determine industry categorisation. Not unsurprisingly, while this has produced some challenges on the issue of 'industry segmentation', the overall result has been one that is, with each successive annual survey, becoming 'more rounded' and inclusive of all industry sectors.

Further research into this issue was undertaken for the 2005/06 survey, which requested information about the skills and professional backgrounds of the SI users within the organisations, with contributors being able to provide four categorisations of the skills in their responses, each with different percentage attribution.

The outcome of the analysis of the skill profile of SI users is shown in the following charts for all industry sectors.

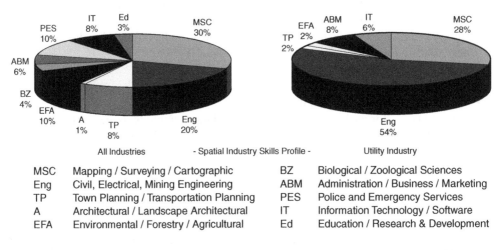

MSC	Mapping / Surveying / Cartographic	BZ	Biological / Zoological Sciences
Eng	Civil, Electrical, Mining Engineering	ABM	Administration / Business / Marketing
TP	Town Planning / Transportation Planning	PES	Police and Emergency Services
A	Architectural / Landscape Architectural	IT	Information Technology / Software
EFA	Environmental / Forestry / Agricultural	Ed	Education / Research & Development

It is clear from these charts that although the Mapping/Surveying/Cartographic component of the SI industry is important, it represents less than a third of the industry (at only 30%), but the Engineering sector is a major participant of the overall industry (at 20%) and is the dominant group of participants in the Utilities sector, at over half of the industry. These statistics provide an insight into the skill background and context required to support the growth of the SI industry.

2. How is 'the value' defined?

   Is it:

   a. the sum of the incomes of companies who are suppliers to the industry, with this information being sourced from suppliers to the industry (i.e. the vendors, consultants, data providers, etc.); or

   b. the sum of the 'spend' of all organisations who purchase this technology, with this information being sourced from end-users in the industry (i.e. government, utilities, private organisations who are bona fide end-users of the technology).

There are a number of problems with adopting option (a):

- in my experience suppliers are (usually) fiercely competitive and are very reluctant to provide any financial information which they think may be accessed by their competitor – and so generally any information that they do provide (if they provide any at all) has to be considered as being 'not the full story', or that it has been 'embellished';
- suppliers are (often) multi-national – therefore any product development, particularly software, which is developed in the UK, USA, Europe or Australia typically contains a large component of work generated in another country, thereby making it extremely difficult to separate out the 'value' to be applied locally (to whichever country that this applies to); and
- only the suppliers who are known to be in the industry are targeted for response, therefore only a component of the industry is asked to provide information, thus leading back to the questions about industry definition (point 1 above).

Option (b) avoids these obvious issues, but relies on having a wide distribution of the survey to all organisations that could possibly use spatial information technology. Option (b) is used to derive 'the values' of the industry in the Australian and New Zealand GIS surveys.

Therefore, the definitions used for 'the industry' and 'industry value' in the GIS Survey are as outlined above, recognising that while everyone will not agree with these definitions, they do provide a practical definition in lieu of no other definitions being available, or found to be workable.

The 2006 Geospatial Technology Report[3] by GITA (refer to http://www.gita.org) describes the State of the Industry based upon 386 survey responses from Utilities and Public Sector organisations categorised by:

- Electric Utility Industry
- Gas Utility Industry
- Pipeline Industry
- Public Sector
- Telecommunications Industry
- Water, Wastewater and Stormwater Industry

While these two reports (the GITA North American report and the Corporate GIS Consultants Australia/New Zealand report) have a different focus, different industry categorisation and are based on statistics from different countries, much of the information reported in the two surveys is reasonably similar, given that each has local conditions.

Each year in the Australian/New Zealand survey there is always an interesting mix of organisations that contribute and, because there is approximately a 40% 'churn' in contributions each year, there is always a reasonably good level of new responses to provide challenging information, particularly on budgets. However, because budget information is often sensitive or difficult to obtain, the section on budget questions typically has the lowest response rate, noting that some contributors do not answer all sections or all questions in all survey sections that they do answer.

---

[3] 2006 Geospatial Technology Report, Geospatial Information and Technology Association (GITA), North America.

In 2006/07, the 'value' of the SI industry for Australia and New Zealand was determined to be approximately 1 billion dollars based on reported budgets across all industries. This is correlated against a year 2000 industry value base of 800 million dollars based on statistics available at that time and from extensive external interviews and investigations undertaken within major segments of the industry.

While the value of the SI industry worldwide is unknown, it is estimated that the value of the North American spatial information industry (USA, Canada and Mexico) is 30–50 times that of Australia/New Zealand, thus suggesting that the North American market is of the order of 30–50 billion dollars, however this is an extrapolation which should be treated with a substantial degree of caution. The information presented in the GITA Geospatial Technology Report tends to indicate that this 'value' estimation is not unreasonable.

The outcome of the 7 years of statistics is that the value of the SI industry in Australia and New Zealand has grown steadily at an average rate of between 5% and 6% p.a. However, at times during these years, the value has grown at up to 11% p.a. This has flattened since 2004/05 with a 21% drop in budgets, indicating that either industry momentum was slowing or that spending during the period 2002–2004 was unduly inflated. Further research indicates that this could have resulted from increased spending on GIS and IT since mid-2002 as a result of the post '9/11 war on terror'.

The segmentation of budgets (for all industry sectors) is shown in the following chart for the budget categories covered in these surveys. These averaged reported budgets for all industry sectors have iterated to being almost constant over successive years.

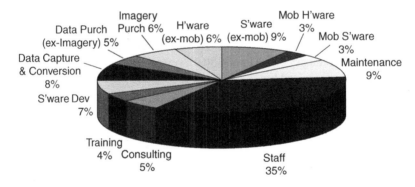

Another useful indicator in the industry is the average budget per site (across all industry sectors), which has levelled off at just over $ 550,000 per site in the 2006/07 survey. The budget trends for the last 5 years on an 'all industries' basis are shown in the table below.

	2002	2003	2004	2005	2006
Staffing	50%	43%	34%	35%	37%
Consulting	5%	5%	6%	5%	5%
Training	4%	4%	4%	4%	4%
Software	6%	10%	13%	12%	12%
Software Development	7%	5%	5%	7%	6%
Data Capture	10%	8%	9%	8%	8%
Data Purchase	4%	11%	13%	11%	13%
Hardware	4%	7%	6%	9%	7%
Maintenance	10%	7%	10%	9%	8%

This 5-year table highlights some interesting trends across all industries:

- there is a very sharp budget reduction from 50% to 37% for staffing over the last 5 years;
- training continues at a very consistent low 4% of total budget spend;
- the spend on software has doubled over the last 5 years;
- the total spend on data (capture and purchase) has risen from 14% to 21% of budget (noting that this is for external contracts and excludes internal staffing costs relating to data); and
- while data capture has remained reasonably static, the amount spent on purchasing data has risen sharply over the 5 years, tripling from 4% of total budget to 13%.

Further dissection of this 'all industries' budget data can be categorised to show the percentage budget splits for each industry sector reported, leading to the table shown overleaf.

From a review of this table, it is apparent that the averages are reasonably consistent across most sectors, with the following highlights:

- the average for hardware spending is 5% of budgets;
- the sector with the largest software spend is Education, which is surprising when one considers that software is often provided free of charge (or heavily discounted) to educational institutions;
- mobile software averages only 2% of overall budgets, with private companies spending the most at 4% of overall budgets;
- maintenance averages at 7%, even though some sectors pay up to 10%;
- staffing ranges from 30% of total budget (for Utilities and GOC) to 54% of total budget (for NFP), suggesting that the latter sector is the place to work, followed by the Federal Government and the Education sector;
- consulting averages at 9% but ranges from a high of 18% for the Federal Government (not surprising) to a low of 3% for the NFP sector (again, not surprising);
- training averages at a consistent 4% almost universally across all industry sectors;
- software development, at 5%, is reasonably consistent across most sectors, but is the highest in Utilities, State Government and GOC sectors;
- data capture/conversion averages at 10%, the highest being GOC at 18% and Utilities at 13%;
- data purchase averages at 8%, with the Federal Government being the biggest purchaser (at 11%) and Local Government and Utilities being the smallest purchasers; and
- the purchase of image products averages 6% of the total budget spend.

2.3 GIS PRODUCT USAGE

There are 14 identified GIS products which have been consistently reported over the last 7 years of the GIS Survey in the Australian and New Zealand SI industry, as shown in the following chart. Of these, nine GIS products were reported as being above 2% of the contributor base.

	Education	Federal Government	State Government	Local Government	Utility	GOC*	NFP*	Private Companies	Average
Hardware	6%	3%	7%	6%	5%	3%	3%	7%	5%
Software (ex Mobile)	12%	6%	7%	10%	10%	7%	5%	10%	9%
Software Mobile	2%	1%	3%	2%	3%	1%	0%	4%	2%
Maintenance	5%	6%	7%	10%	10%	6%	9%	5%	7%
Staff	40%	40%	35%	32%	30%	30%	54%	33%	37%
Consulting	10%	18%	5%	6%	8%	11%	3%	9%	9%
Training	3%	4%	4%	4%	4%	4%	3%	5%	4%
Software Development	3%	5%	7%	5%	7%	7%	0%	4%	5%
Data Capture/Conversion	9%	4%	9%	9%	13%	18%	9%	7%	10%
Data Purchase	6%	11%	9%	5%	6%	7%	9%	8%	8%
Imagery Purchase	3%	4%	7%	9%	5%	6%	3%	8%	6%
Forecast Next Year Budgets	**+8%**	**+11%**	**+2%**	**+9%**	**+8%**	**−3%**	**+3%**	**+15%**	

* GOC, government-owned corporations; NFP, Not for Profit organisations.

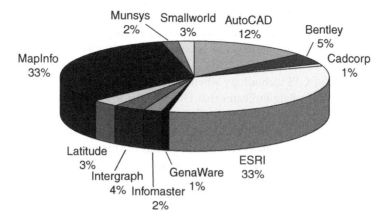

In the 2006/07 survey, there were 878 instances of GIS vendor product usage reported by the 494 contributors. This averages to 1.8 vendor products per site, which is slightly higher than previous averages of 1.7 vendor products per site from previous years.

In reviewing these statistics, please note that this information is based on *reported* use of GIS technology from contributors and that while software suppliers and other vendors were not requested to supply customer contact information, they were all invited to forward a request to complete the survey to their customers.

As such this information should *not* be used as an indicator of product 'market share' – it is information reported on a voluntarily basis only. Nevertheless, this data has been found to be a reasonable indicator of some components of the market.

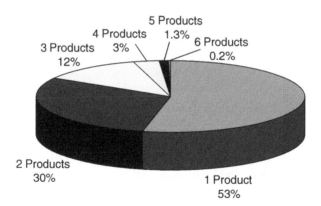

In the Australia/New Zealand survey, almost half of all GIS users have two or more GISs, with a small percentage having up to six GISs per organisation, whereas the North American Survey[4] shows that a massive 70% of organisations used two or more GISs.

Based upon these reported statistics and in keeping with similar trends from previous years, it is apparent that MapInfo and ESRI have the highest reported GIS product usage in

[4] 2006 Geospatial Technology Report, Geospatial Information and Technology Association (GITA), North America.

the Australian and New Zealand market based on the contributors to this survey. However, the results of this survey indicate that while the 'major' suppliers are further entrenched into the market, some of the markets are becoming more fragmented. This is nowhere more evident than in the Utility sector where there continues to be a substantial change reported in suppliers of GIS technology when compared to that of previous years.

Further analysis of this data indicates that GIS product usage across the Federal Governments in Australia and New Zealand reflects that USA software products ESRI and MapInfo hold the dominant role, with AutoDesk gaining increased product use. However, while Local Government shows that MapInfo has the highest reported GIS usage at 35%, followed by ESRI at 24%, they do support a number of local GIS software products. Utilities report that the use of MapInfo is at 23%, ESRI at 18% and Smallworld at 14%. Again, the lack of product loyalty (i.e. multiple product use) is endemic in the SI industry.

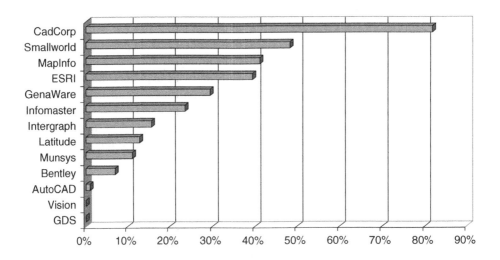

The above chart shows a comparison between the reported percentage use of one versus more than one GIS product, categorised by vendor. In each case the percentage use of more than one GIS product is the difference between the bar percentage and 100%, e.g. ESRI customers reported 39% usage of one product (i.e. ESRI) and 61% usage of non-ESRI products. CadCorp and Smallworld customers are the most loyal, reporting the highest incidence of single product set use (at 82% and 48%, respectively). Interestingly, the two product sets in most prolific use, MapInfo and ESRI, are reported as third and fourth each with 41% and 39% single product set usage, respectively. This reported lack of product loyalty, combined with the churn of product purchases, suggests that vendors may need to refocus on the customers and/or provide better service to a GIS industry which is becoming more cynical and less loyal to specific products.

For those respondents who indicated that they had a previous GIS, the distribution of prior systems was reported as shown in the following chart.

As can be seen from this chart, almost no Vendor has escaped from having their system being replaced by another. GenaMap was reported to have passed GDS as the vendor from which most users are moving away from.

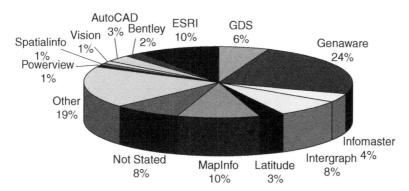

The change of GIS products (churn) in the SI industry continues to be a major factor. In the last 4 years, the product churn has fallen from 48% of GISs being replaced in the previous 3 years to currently being 30% of GISs being replaced in the previous 3 years. While this may seem to be an improvement, the early figures are less certain due to a poorer statistical base in the early years of the survey. Nevertheless, this level of product churn must be considered to be an alarming figure, particularly for vendors who continually have customers moving to alternative platforms.

The major reasons for this churn were:

- the GIS did not meet current needs – 36%;
- the GIS did not conform to a common strategy – 14%;
- the GIS was a superseded product – 13%.

It is clear that the 'churn' in the market is a major factor which highlights a continuing level of dissatisfaction with vendor performance and GIS capabilities. Perhaps vendors should pay closer attention to the concerns of their user community and undertake some research into why their customers are moving on to other products.

2.4 SPATIAL APPLICATIONS

The application uses of GIS can be a useful indicator of the degree to which the capabilities of this technology are currently being harnessed for daily business use. Each year there are generally about 3500 application units reported for the survey across the 27 application areas, distributed as shown in the following chart.

The 'top three' applications used each year are consistently reported as being Cartographic Mapping, Asset Management and Cadastral Management. Interestingly a number of business-based applications such as Customer Management, Agricultural and Economic Modelling are substantially 'down the list' of priorities. In addition, contributors reported on whether the functionality used was Off-the-shelf; Off-the-shelf but customised; or Custom application. The sophistication of applications surveyed was categorised as: Simple (i.e. visually inspect datasets and issue simple queries only); Moderate (i.e. use a cross-section of advanced functions to compose and analyse datasets); Sophisticated (i.e. business rules are embedded to assist in composing and analysing datasets); and Complex

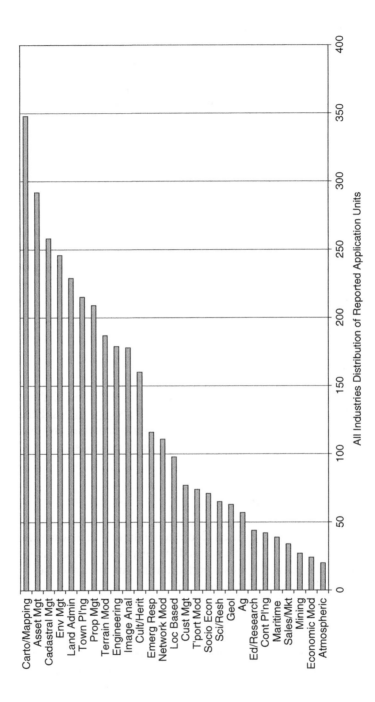

All Industries Distribution of Reported Application Units

(i.e. business rules model real-world processes and workflow, i.e. much of the composition and analysis of datasets is automated).

The results of this analysis showed that most GIS use was spread between customised and non-customised applications and that generally they were not too complex. The major uses of this technology are focused on:

- Asset Management for Local Government and Utilities organisations; and
- Cadastral Mapping applications for all other sectors.

This usage tends to indicate that the use of spatial technologies is still at a 'grass roots' level and is focused on day-to-day tasks. Application use for Image Analysis, Engineering Analysis and Terrain Modelling also rate highly, indicating that users are exploring areas which 'add value' and focus on specific business needs, but they are not doing so to any significant extent.

Location-Based Services (LBS) and Customer Management rated 'average' in all sectors except Utilities, where there is a focus on making the most use of technology to address business needs.

2.5   TRAINING

The training being undertaken in the SI industry is a useful indicator of the available skills and their distribution across industry sectors, indicating areas where more resources need to be spent. This is an area which should be of particular interest to the industry, particularly when staff budgets are falling and training remains at a low 4% of overall budget. The training needs for the following course types were canvassed, noting that while some of the advanced training is obviously the providence of specific vendors, the basic level training is generic and (in most cases) vendor independent.

1. Basic User Training – introductory GIS use.
2. Advanced User Training – advanced GIS use, GIS customisation and introductory SQL server.
3. Designer Training – data modelling and database design.
4. Developer Training – introductory and advanced GIS programming, programming with spatial database engines and extensions and web application development.
5. Specific Technology Training – Global Positioning System (GPS)/GIS integration, image analysis and data capture using mobile computing software and techniques.
6. Project Management Training – application/system development and project management methodologies.

The following chart shows the level of current and future training units (right-hand bar) reported for all industries for the last 5 years for Australia and New Zealand. It indicates a healthy growth of approximately 13.5% for numbers of current training levels and approximately 12% for numbers of future training requirements. It also indicates that there is a substantially large and trained user base available to meet industry requirements. However, this is of particular interest when considered in conjunction with:

- the budget value of the SI industry – growing at 5–6% p.a. for the last 6 years;

- the level of system churn (thereby requiring re-skilling) – currently at 40% over the last 3 years (approximately 18% p.a.); and
- the percentage of budgets spent on staff – fallen from 50% to 35% over the last 4 years.

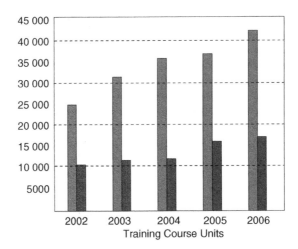

Training Course Units

The following chart shows that there is a high level of staff in the industry who are trained at an introductory level as well as a high level of introductory training required. When this is correlated with responses to system churn and questions on product deployment methodologies, it is apparent that this high requirement for introductory training is most likely due to the high and continual 'churn' of new people needing this introductory training across a number of government and utility organisations.

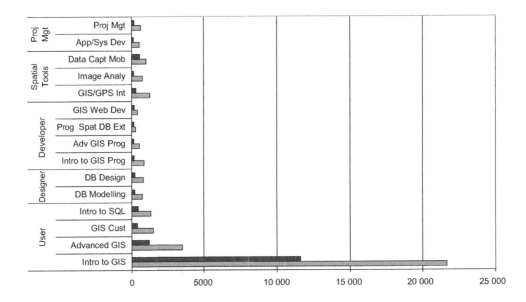

Against this backdrop of high industry growth and reducing staff base with a static and low level of budget spent on training, the unfulfilled training requirement remains high. This training gap has led some user organisations to report that they have designed and implemented their own suite of courses, in some cases with the expectation of gaining certification.

Further work was undertaken to determine whether there was a skill shortage in Australia and New Zealand, in particular to identify training deficiencies and to understand how an appropriate level of training could be provided in a context-sensitive and effective manner to meet current and expected industry growth.

In summary, the issues associated with training/skilling in the SI industry are:

- The number of people trained each year is growing at approximately 14%, notwith-standing a smaller rise in 2005.
- Almost half of the industry does not have a skills shortage and of those that have re-ported having a shortage, only a third consider that this can be addressed by additional training.
- Most of the training requirement is at the introductory level and is focused on vendor-specific and organisational specific training.
- Half of the industry has developed and is running in-house courses, again at the introductory level and focused on organisational-specific training and vendor-specific training.
- While the industry value is growing, training budgets remain static at 4%.
- Staff churn at the GIS Officer level is between 3 and 5 years, further adding to the need for more Introductory Training to be undertaken.
- System churn continues to be at a high rate, in some instances up to 18%, further exacerbating the need to re-train staff for changed technology implementations.
- The staff profile of the industry indicates that only 30% are from a Mapping/Surveying background, yet the focus of most training in tertiary educational facilities is inap-propriately directed at these disciplines.

2.6   SPATIAL DATA

Spatial Systems must have adequate data in order for business requirements to be met. Spatial data continues to be a complex issue and one which a number of government agencies have invested considerable effort to address. Of particular concern is whether the data is 'fit for the purpose' to which it is intended to be used and whether the accuracy, quality, completeness and currency are to the level required.

The GIS Survey has compiled substantial information about the nature and use of data themes over the last 7 years, with the total number of datasets reported as being used continuing to grow at between 5% and 7% each year. The following chart shows the ranking of data usage on an 'all industries' basis across all geographic areas. While the most used datasets are those traditionally provided by government agencies, e.g. cadastre, topographic, land and titling, administrative boundaries, etc., some of this data, such as aerial photography and road data, is also provided by the private sector in some circumstances. This data can then be reviewed on an industry sector basis, ranked by the

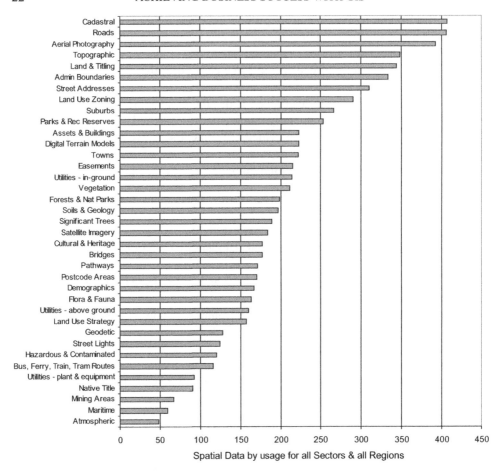

Spatial Data by usage for all Sectors & all Regions

reported importance of that dataset, the amount of that data which is sourced internally and the fitness for purpose of that data, as follows.

DATA USED BY FEDERAL AUSTRALIA AND NEW ZEALAND GOVERNMENTS

The top ten datasets used by Federal Australian and New Zealand Government agencies are shown in the table below, ranked by a categorisation of their importance (to their agency), the percentage of the data which is sourced internally and the fitness for purpose of that data. Topographic data, terrain models and imagery (satellite and aerial) are the datasets most used by Federal Government, obviously for broad planning and regional issues. The Federal Government (both in Australia and New Zealand) rate the cadastre as being of a lower priority than any other industry sector, indicating that this level of data granularity is not required as much for government at this level.

Very little Federal Government data is sourced internally but it has a moderately good level of fitness for the purpose.

Top 10 Federal Government Databases	Importance	Data Sourced Internally	Fitness for Purpose
1. Topographic	54%	17%	49%
2. Satellite Imagery	49%	13%	49%
3. Digital Terrain Models	46%	27%	39%
4. Aerial Photography	44%	11%	49%
5. Roads	42%	13%	54%
6. Maritime	38%	36%	36%
7. Towns/Places of Interest	34%	15%	33%
8. Vegetation Boundaries	32%	26%	33%
9. Cadastre	29%	0%	30%
10. Demographic Data	29%	7%	24%

DATA USED BY STATE GOVERNMENT

The top ten datasets used by State Government agencies are shown in the table below. In some, but not all, cases there is a correlation between the datasets which are sourced and maintained internally with the fitness for purpose of that dataset. For example, Street Addresses is 0% sourced internally (since it is derived from Local Government) and has a correspondingly low level of fitness for State Government purpose.

Top 10 State Government Datasets	Importance	Data Sourced Internally	Fitness for Purpose
1. Cadastre	70%	14%	54%
2. Aerial Photography	68%	0%	71%
3. Administration Boundaries	61%	12%	59%
4. Street Addresses	57%	0%	20%
5. Land Title/Ownership	55%	20%	51%
6. Topographic	52%	17%	51%
7. Forests/National Parks	52%	9%	56%
8. Roads	50%	20%	44%
9. Parks and Recreation Areas	46%	20%	49%
10. Hazardous/Contaminated Sites	43%	50%	29%

The data used by State Governments is mainly focused on the cadastre, aerial photography and datasets which are used for administrative purposes. Remotely sensed satellite imagery is not in the top ten of the State Government's data requirements. Surprisingly, the cadastre is reported as being only 70% sourced internally even though it is maintained by (almost) all State Government agencies, noting that these statistics are for all State Government agencies (not just the Lands Agency) and for all regions in Australia and New Zealand.

DATA USED BY LOCAL GOVERNMENT

The top ten datasets used by Local Governments are shown in the table below. Interestingly, even some data which is sourced internally (e.g. street addresses, land use zoning plans, etc.) still do not have a high degree of fitness for purpose. For example, Parks and Recreation data is 72% sourced internally but is only 6% fit for that purpose.

Top 10 Local Government Datasets	Importance	Data Sourced Internally	Fitness for Purpose
1. Cadastre	99%	40%	29%
2. Land Title/Ownership	92%	49%	32%
3. Land Use Zoning Plans	91%	94%	36%
4. Roads	86%	56%	27%
5. Street Addresses	80%	92%	31%
6. Aerial Photography	75%	2%	30%
7. Suburbs	63%	44%	22%
8. Administration Boundaries	60%	36%	25%
9. Parks/Recreation Areas	55%	72%	6%
10. Topographic	50%	16%	17%

As would be expected, the data used by Local Governments is focused on property, ownership and zoning followed by all the other datasets which are generally used for administrative purposes. In most jurisdictions, the cadastre is sourced from the government agency responsible for the maintenance of the cadastre (usually the Lands Agency) in that state/region.

Case example

The Australian Spatial Data Directory (ASDD) is an online directory that enables the discovery of spatial data which is available throughout Australia. The information contained in the directory is metadata, a summary of information about the dataset, including the geographic area that the dataset covers, the custodian, who to contact to obtain a copy of the dataset and other useful information that helps people decide whether or not the dataset is useful for their particular purpose. The ASDD was launched in 1998 and now contains over 50,000 entries held on 24 nodes around Australia. The ASDD allows the concurrent interrogation of the existing nodes by a user with an internet browser.

While awareness of the ASDD is increasing, only 40% of GIS users make use of it to discover data and less than 20% of GIS users add data to it, no doubt due in part to the poor functionality reported. Further detailed analysis indicates that ASDD uptake is different in different jurisdictions and industries.

- the Australian Capital Territory has been the most successful in adopting ASDD principles, followed by Western Australia, Queensland and NSW. Other states such as Victoria, Tasmania and South Australia are poor users of the ASDD.
- the Australian Federal Government, followed by Education and State Government, are the most effective sectors using the ASDD. Utilities, Local Government and the Mining Industry almost never use the ASDD.

In a number of jurisdictions, the cadastre is maintained by a State Government agency as well as by numerous Local Governments and, in some cases, Utilities. In New South Wales for example, over 90% of Local Governments maintain their own cadastres. While Councils may have received their cadastres from the Lands Department or from the major water utility Sydney Water (some over 10 years ago), only 10% of Councils regularly use the Lands cadastre for undertaking Council transactions – 90% of Councils who receive annual updates of the Lands cadastre report that they use this data only for checking discrepancies against their own cadastres and then continue to use their own cadastres for transacting Council business.

This has been a longstanding and major issue for GIS users in NSW, particularly in the Sydney metropolitan region. It is estimated that there are in excess of 20 cadastres being maintained by Councils, Utilities and government agencies in the Sydney area, mostly to separate spatial and attribute content and spatial positioning specifications based on individual business rules and for the geographic area of responsibility of each organisation.

While some of the cadastres used by Councils have been derived from the Sydney Water cadastre at some time in the past, they have often been maintained separately and have therefore diverged from that maintained by Sydney Water. In addition, the Lands Department has for some time provided a cadastre for the Sydney area based upon data captured through the land registration process, again captured differently to that by Sydney Water. The current Single Land Cadastre project, a joint initiative of the Lands Department and Sydney Water, should result in a 'harmonised' cadastre being developed, with the cadastre for the remaining part of the state being similarly progressively upgraded/updated to take into account those Local Government cadastres in regional areas.

DATA USED BY UTILITIES

The top ten datasets used by Power, Water, Gas and Telecommunications Utilities are shown in the table below. Utilities are the only industry sector who universally rate the cadastre as being 100% important to their business operations, unlike State Governments who rate the cadastre as being only 70% important to their business. In-ground utility data is (as expected) the next important dataset with a very high level (91%) of fitness for purpose, even though only 65% of this data is sourced internally.

Top 10 Utility Datasets	Importance	Data Sourced Internally	Fitness for Purpose
1. Cadastre	100%	27%	62%
2. Utility data (in-ground)	95%	65%	91%
3. Roads	77%	20%	60%
4. Street Addresses	73%	33%	42%
5. Assets	68%	100%	34%
6. Easements	68%	50%	36%
7. Land Title/Ownership	66%	50%	51%
8. Suburbs	66%	30%	63%
9. Utility data (above ground)	64%	100%	49%
10. Administration Boundaries	61%	0%	49%

Surprisingly Assets data, often 'core business data' for Utilities, is rated fifth at only 68% important, even though this data is 100% sourced internally (as expected), but is only 34% fit for purpose. It seems that some utilities should review the data which they are collecting for assets and endeavour to make this data more appropriate to meet their business needs.

DATA USED BY THE EDUCATION SECTOR

The Education Sector report that they generally use 'whatever data is available' and often that data is provided on a 'free' or 'demonstration only' basis from data providers, including government agencies, on the understanding that it is not used for any commercial purposes.

DATA USED BY THE PRIVATE SECTOR

The data used by the Private Sector can be categorised into two components:

- the 'provider' component will capture and use data on a 'project-by-project' basis for whatever data the client wants – as such, there are no specific generic 'industry' datasets for this component; and
- the 'user' component is focused on the data relevant to the business – as such, there are no specific generic 'industry' datasets for this component.

2.7 IMAGERY

Imagery (satellite and airborne) is a very useful dataset in SI environments, either:

- in a passive role as a qualitative 'backdrop' to vector data, i.e. a landscape onto which vector data can be applied to provide contextual relationships (e.g. to show the vegetation which will be impacted by a reclamation zone, etc.); or
- in an active role as the basis for mathematical analysis of the spectral bands to highlight specific features (e.g. diseased crops on a landscape of wheat, saline soil, etc.).

Over 85% of contributors to the GIS Survey consistently report that they use imagery as an integral part of their SI environment. The use of imagery across each industry sector, normalised for numbers of responses per sector, is as shown in the following chart, showing those who do use imagery (bottom bar) and those who do not use imagery (top bar) in each of the sectors.

While most industry sectors make substantial use of imagery in their GIS environments, the biggest users are Local Government, State Government and government-owned corporations (GOC).

The business use of imagery is shown in the following chart. Nine specific uses were reported over the last 3 years across a total of over 1,500 units of use reported, increasing by 15% over this time.

Mapping and planning were the most prolific uses reported, collectively accounting for 60% of use, followed by engineering and natural resource management. Again, those who

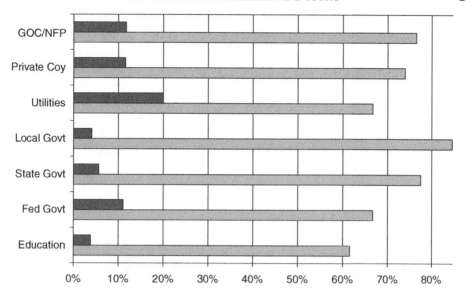

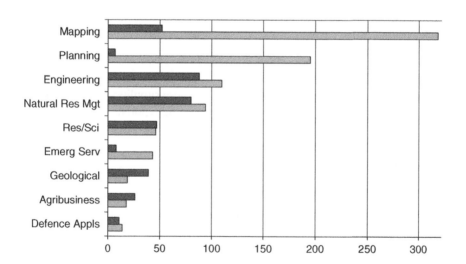

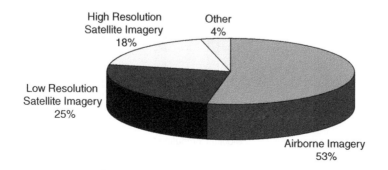

do use imagery are shown by the bottom bar against each business application and those who do not use imagery are shown by the top bar.

The imagery product types used are as shown in the previous chart, indicating that airborne imagery clearly provides the majority of products used in the SI environment, being consistent with the majority of business uses (shown above) of mapping and planning.

The following chart on the left shows the breakdown of airborne imagery into specific product types, with photography clearly being the highest use of airborne imagery. The chart on the right shows the breakdown of the satellite product units, indicating that multispectral is the dominant satellite product used, followed by panchromatic at a constant 26% for the last couple of years.

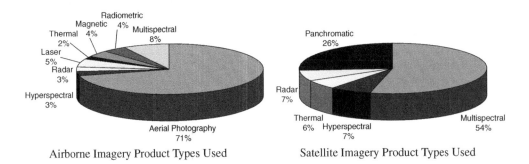

Airborne Imagery Product Types Used   Satellite Imagery Product Types Used

The supply of imagery product by the private sector was reported across 238 product purchase units in the latest survey over 35 suppliers. These suppliers, and further analysis of the provision of imagery, are discussed in other specific reports.

2.8   MOBILE COMPUTING

The numbers of mobile devices in the SI industry are increasing substantially each year as shown in the following chart, with 77% responding in 2006 that they use mobile technology.

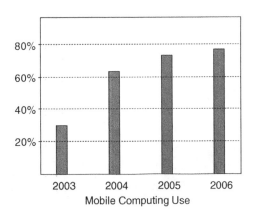

Mobile Computing Use

ESRI and MapInfo products account for 81% of the reported software product usage on mobile computing devices, which is up approximately 15% each year over the last 3 years. The cost of software for mobile devices was considered to be reasonable by a third of the contributors, while another third considered that the cost was significant.

It is clear that the mobile computing market is undergoing substantial growth and, although the numbers of software products are increasing, this market segment remains dominated by two vendors. Note that this is reported usage only and is not indicative of market share.

The broad business purposes for which mobile computing is used across six specific business functions are indicated in the following chart, showing the growth over the previous year.

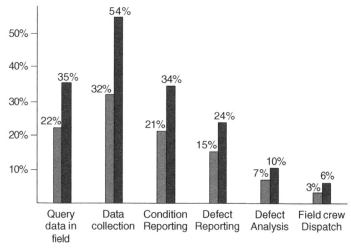

Business Use of Mobile Computing

The most prolific growth and use is in data collection (for input/update of office-based systems) and query access to data while in the field. The types of mobile devices used (in reasonably similar proportions) are Laptop, PDA, Pen Tablet and Smart Phone across a standard configuration, ruggerdised and/or a mixture.

2.9 REGIONAL SI INITIATIVES

Most jurisdictions in Australia and New Zealand have developed comprehensive multi-agency SI strategies, policies and implementation processes over a number of years; in some cases this has been progressing for a long period of time – 25 years in the case of WALIS, the Western Australian initiative.

All SI strategies are similar in context and subscribe to good governance and standard principles, with considerable experience, policies and structure being accumulated over that time by all jurisdictions. In almost all cases, there is a coordinating Council or Committee with representation from the major participating (data) agencies and councils and, with the exception of Queensland, this strategy is managed by the 'Lands' agency in that jurisdiction.

Information about policies and initiatives for specific jurisdictions can be accessed at the following web sites:

- Australian Capital Territory
  http://www.urbanservices.act.gov.au/libinfo/geoinfomgmnt/gim.html
- New South Wales
  http://www.bossi.nsw.gov.au/moreinfo/notices.html
- Northern Territory
  http://www.ntlis.nt.gov.au
- Queensland
  http://www.qsic.qld.gov.au
- South Australia
  http://www.environment.sa.gov.au
- Tasmania
  http://www.thelist.tas.gov.au
- Victoria
  http://www.land.vic.gov.au
- Western Australia
  http://www.walis.wa.gov.au
- New Zealand
  http://www.linz.govt.nz

2.10   SUMMARY

In summary, the GIS/SI industry in Australia and New Zealand is significant, fast-growing and complex. This industry is dominated by Government (at all levels), Utilities and quasi-government, with the private sector generally being a provider to this government-dominated industry.

The 2006 Geospatial Technology Report, developed by GITA North America, also reports similar observations, albeit in a utility-by-utility comparison. And, as often happens in a fast-growing and complex industry, there is considerable scope for confusion about how one should apply options available for technologies, standards, implementation considerations and data capture/conversion programs.

While initially most users of SI technology have 'grown' with the technology, often at the behest of the technology provider they have used, it is apparent that users are now becoming more sophisticated and have come to understand that there is a need to have an overall strategy for the use of this technology to meet their business needs. As users 'churn' through successive technologies, they are also requiring that their strategy should not be based on that of their technology provider.

In the last decade, most SI users have become more empowered, generally changing their GIS product until they find one which meets their business needs. As such, it is becoming more apparent that 'getting it right' (in a business sense) is becoming more important than the technology.

One of the best methods of 'getting it right' is to develop a strategy for GIS based on the business of the organisation so that there is a 'roadmap' which can be followed when the organisation embarks on this journey.

Further measures include measuring the performance of the host organisation against that of other like organisations. An outcome of these external comparisons can be used to show areas that may need further improvement to increase efficiency and/or reduce cost. Benchmarking Studies (discussed further in Chapter 11) are a useful method to undertake these external comparisons, particularly for groups of like organisations (e.g. Utilities, Councils).

As such, these processes can also be used to further refine 'Best Practice' for GIS in each industry sector. The application of this level of 'Best Practice' can then be used to improve the effectiveness of GIS to meet the business requirements for the host organisation.

3 Introducing the Elements of a GIS Strategy

The outcomes from the GIS/Spatial Best Practice Survey over the last 7 years highlight that two-thirds of organisations using GIS do *not* have a strategy and that over 70% do *not* use Key Performance Indicators (KPIs). That is, they have *no* destination, *no* roadmap and *no* possible way to measure and record success. Is it any wonder that over 50% of GIS managers then go on to report that their GIS is not widely understood across their organisation. How can it be understood by the wider organisation, when the goal of it (the GIS) appears not to be understood by the GIS Managers themselves?

Given these statistics, it is clear why a large number of GIS projects either fail or are less successful than they otherwise should have been. That is because there is no plan and no understanding of what constitutes the goalposts and where they are. And because the senior managers also do not know where the goalposts are, nobody knows when a goal has been kicked.

Example

Imagine a game of football where there were no goalposts:

- Would the players know which way to run?
- Would the players run in different directions?
- Would the players ever score a goal?
- Would there be any point to the game?
- Would fans come to watch?
- Would the coach say the players had done a good job?
- Would the players get paid or get a performance bonus?
- Would the club get any sponsorship or future support?

The answer to most of these questions is obviously no, but this analogy can be applied to over half of all GIS sites. This is one of the major reasons why GIS managers do not have the support of their senior management and why the role of GIS is not understood by key organisation staff.

Clearly, if these 70% of GIS managers had a strategy and understood what they were trying to achieve and how to measure it, they might have been able to communicate that to the rest of the organisation and their managers, so that everyone could understand what they were doing and would know when they scored a goal.

Achieving Business Success with GIS   Bruce Douglas
© 2008 John Wiley & Sons, Ltd   ISBN: 978-0-470-72724-9

Developing a GIS Strategy is essential to set these goalposts and to communicate where the goalposts are to the rest of the organisation, so that when a goal is scored everyone will know that a goal has been kicked and will stand and cheer. Such a strategy must meet the business/statutory needs of the organisation, encompass the data/information that the organisation needs to undertake its business, be organisationally suitable and be technologically (software and hardware) appropriate.

While technology should not drive the business, the rapid improvements in some technologies can provide opportunities for some businesses to gain significant benefits. For example, the use of mobile and field computing, particularly when combined with GIS and GPS, can have the potential to change the way a business operates, particularly when coupled with field workforce management software. Therefore, while the development of a GIS Strategy must be primarily focused on the business needs of the organisation (rather than on the technology itself), there are some technologies which may influence particular business directions.

So how are the technical and business issues tied together in a comprehensive business strategy that makes sensible use of technology without being too focused on it (the technology)? The following sections outline some of the problems found using traditional IT methodologies and proposes a spatial approach to address this issue.

3.1   THE TRADITIONAL IT STRATEGY APPROACH

There are a number of methodologies currently available on the market which provide a comprehensive approach to developing a traditional IT Strategy. Typically, these methodologies are based on an approach which has a focus on the Business, Information, Applications and Technology (BIAT) steps, as follows:

Each step in this process is focused on:

1. Business issues/needs – i.e. understanding the needs of the organisation to meet operational, business and statutory outcomes.
2. Information/Data issues/needs – i.e. determining the information and data required to support the business needs and determining how that can be acquired.
3. Application issues/needs – i.e. understanding the application software required to support the business and information needs. and
4. Technology issues/needs – i.e. determining the underlying infrastructure required to run the applications.

However, the development of a Spatial Strategy is different to the development of a traditional IT Strategy in a number of ways:

a. It is often very difficult for a person from a business unit to state his/her business requirements in the context of a GIS (as they may do with other IT systems) unless he/she has used or seen this type of technology previously. This is particularly the case when GIS is used in a reasonably intelligent manner such as undertaking spatial queries rather than for simplistic mapping.

This lack of comprehension generally does not happen when dealing with traditional IT areas such as Accounting systems or Human Resources systems because these types of technologies have been around for decades and are actively used by many types of organisations. Therefore, I tend to talk to managers about their business, not about GIS. The role of the Business Analyst in this process is to understand the business and then to take the expressed business needs and consider them in the context of how a technology-based strategy may, or may not, assist those business needs to be met.

It is always easy to spot the novice in this process – they are usually the people talking about the need to have an internal champion or the need for the Chief Executive Officer (CEO) to understand how this can be really useful, that is 'we really need to convince him how great this is'. To that I would say 'sorry guys, you have the wrong focus'. It is not relevant that the CEO has to understand how an internal combustion engine works just so that he can haul goods around in his truck. It is the job of the GIS Manager or Business Analyst to understand the business (as viewed by the CEO) and then to interpret the business requirements so that a plan can be developed and presented to the CEO in the business context that the CEO understands so that he can appreciate how it will help his business.

If pressed, I have also found it useful to reference GIS in the context of simple mapping, such as one would find on many phone directory web sites or on Google Earth. Most people have seen these sites and subsequent discussion can be used to explain some of the concepts.

b. Spatial data is different to non-spatial (textual) data. Spatial Information systems are data-centric, and without good spatial data stored correctly the GIS will be less than effective. That is, it should be understood that while:

- spatial *is not* special when discussing technology (generally);
- spatial *is* special when discussing spatial data and the architectures required to manage spatial data corporately.

This is because Spatial/GIS systems usually involve very large volumes of spatial data with complex structures and, because these large volumes of complex data often have an impact on system performance, a set of different approaches often needs to be developed to accommodate the volumes and complexity of the data so that a piece of functionality can be delivered across a network in a timely manner.

In order to address these issues, a genre of middleware software has been developed by some of the major GIS vendors in this industry, termed 'spatial data engines', to specifically enable geo-processing of very large volumes of very complex data in a fast manner. That is, spatial systems have specific data needs, and specific system needs to meet these data needs, which are quite different from IT environments and should be treated as such.

c. Most traditional IT methodologies treat data conversion as a one-off exercise for the transactional data of the system being implemented. However, in GIS environments, the structure of the data and the methodology of the data conversion can be a major and ongoing process which can often take several years to address and which can have major implications on the workability of the GIS when implemented.

d. In developing a GIS Strategy, there is a strong need to focus on organisational issues because spatial data is typically stored in organisational-based information silos and the breaking down of these information silos must address the organisational issues inherent in these silos.

This is important because GIS applications are data-centric and to successfully address business needs the GIS must be able to successfully access and/or amalgamate the data in these business unit silos. Therefore a Spatial Strategy often needs to have a higher focus on organisational issues than that which would be required for an IT Strategy.

e. While the technology layer is important in a traditional IT environment running large single-purpose corporate applications, this technology focus is not as important for GIS since almost all GIS products run on Windows environments using corporate Relational Database Management Systems (RDBMS). In addition, the application layer has less importance in GIS environments given that most GIS solutions will meet most of a customer's needs most of the time – that is, if a GIS Strategy is appropriately focused on the business, organisational and information issues, then a GIS strategy could be just as successful if it were based on any one of the dozen GIS products on the market. Note that most vendors would strongly disagree with this statement because of their need to differentiate their product from the competition. Nevertheless, this issue is discussed more fully later in this book.

All of these issues would suggest that the methodology to be used for the development of a Spatial Strategy should be modified from the methodology used for an IT Strategy to ensure that there is an appropriate focus on the issues which need to be addressed so that the Spatial Strategy will meet the needs of the business.

3.2   THE SI STRATEGY APPROACH

In order to address the issues discussed above and to ensure that there is an appropriate focus on organisational issues, the traditional BIAT approach should be modified to become a BIOA approach, as follows:

That is, the development of a GIS or SI Strategy should be focused on:

1. Business issues/needs – i.e. understanding the needs of the organisation to meet operational, business and statutory outcomes.
2. Information/Data issues/needs – i.e. determining the information and data required to support the business needs and determining how that can be acquired.

3. Organisational issues/needs – i.e. understanding the organisational structure neces-sary to support the business needs so that the GIS Strategy can be implemented.
4. Application and Technology issues/needs – i.e. developing an understanding of how the GIS applications and related technology can meet these business requirements.

The following chart shows the level of importance of each of these steps, noting that a good understanding of the Business issues is very important whereas the determination of the Applications is very low in importance. Unfortunately many novices focus on the latter rather than the former, with the result being that the subsequent strategy does not meet business needs and is typically unsuccessful.

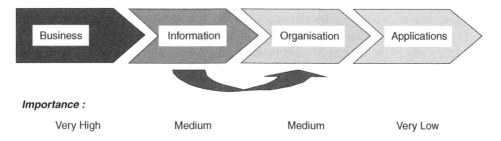

Importance :

| Very High | Medium | Medium | Very Low |

Note that the Information and Organisational steps can be interchanged in this process if required, particularly if the level of organisational-based information silos is high and there is a need to focus on the organisational issues before being able to determine the information and data required to support the business needs.

In summary therefore it should be apparent that the first step, and the most important step, in developing a GIS Strategy is to define the business needs. When the business needs have been determined, *and not before*, the strategy can then address other issues such as data, organisational issues (including training and staffing) and applications/technology issues. In this process, considerable discipline must be exercised to ensure that the application and technology issues are addressed as the final step and not as the first step.

These four elements are discussed in each of the following chapters in some detail prior to developing the GIS Strategy in Chapter 8.

3.3 INFLUENCES OF DISRUPTIVE AND DISTRACTIVE TECHNOLOGY

The IT industry in general, and the GIS industry in particular, can in many cases be rightly criticised for being 'vendor led' rather than 'user driven'. That is, rather than user organisations determining the functionality required to support their business processes, in many cases these organisations are pushed along a specific path by a technology supplier, often the incumbent supplier, so that the customer purchases the technology which the vendor supplies.

This is not always a bad thing – in many cases the direction that a vendor is taking may be that which all of his customers are taking (i.e. the industry collective) and therefore the concept of being in line with industry trends can provide a depth of software func-tionality determined by general industry needs at a collective cost which can be provided to many organisations for a lot less than if that capability were developed for a single customer.

Some of this 'vendor push' technology can also be so different from the mainstream that it can change the way that business may be able to be conducted. Often this is referred to as 'disruptive' technology, some examples of which have the potential to radically change the way that functions are performed:

- The Internet is now commonly used as a vast resource to interconnect people and organisations, to undertake extensive research and to find anything that one may require, and while the Internet is truly a disruptive technology, it was not developed with this in mind, but rather was collectively developed by the United States Federal Government several decades ago as a means for researchers to communicate.
- Google Earth has totally changed the way that people engage with mapping and SI, particularly in the context of downloading data in real-time from the Google server as well as for creating SI environments based almost entirely on image and not vector landbases.
- Mobile computing devices such as small Laptops and Personal Data Assistants (PDAs) have provided the ability to gather information and undertake computing while on the move, albeit sometimes at a slower speed, in many cases transforming the way that business is transacted.
- Storing spatial data in rows and columns in a relational database such as Oracle Spatial means that the spatial data can be divorced from the spatial applications which are used to edit and create that data. In this manner, different GIS products can access a single database of SI in real-time or near real-time, depending on the level of transaction management provided. This has the potential to provide the ability to exchange data at the database level and, in so doing, to enable use of a 'best of breed' approach of different GISs focused on different business applications. This approach has also negated the need to move data from one GIS to another, in that it is stored in an independent database. This is very significant technology and one which has the potential to make a substantial difference in the industry.
- Web Services[1] allows any piece of software to communicate with a standardised XML (eXtensible Markup Language) messaging system, with benefits including platform independence, ability to exchange data between different businesses, ability to choose the best technology platform for each situation and location and device independence.
- The extension of this to Web Mapping Service/Web Feature Service (WMS/WFS) specifications, based on work undertaken by the Open Geospatial Consortium (OGC), provides the ability for organisations to exchange data, or to access data served by another organisation using WMS or WFS. However while this is an excellent initiative which promises useful business outcomes, and is promoted by some as a panacea, it is not yet apparent whether WMS/WFS will proceed to the mass implementation stage necessary to become truly disruptive.

So when developing a strategy, it is important to recognise and take advantage of relevant genres of technology which may be able to provide radically new business benefits that may not have been otherwise realisable.

---

[1] A good explanation of Web Services can be found at http://en.wikipedia.org/wiki/Web_service.

And this is important because, all too often, user requirements are determined from a baseline of knowledge of the user, and if the user has a poor understanding of the technology available on the market and that which can be achieved with this technology, then that person may have set their expectations at a particularly low level, or in the wrong direction, and not take advantage of capabilities from new technologies which may be useful to the business.

In this manner there is a dichotomy between 'vendor push' and 'user pull' technology which should be understood so that one can take advantage of relevant technology to meet business needs where applicable and not develop a strategy by looking at where one has been, rather than where one should be going. The analogy about not trying to drive by looking in the rear-view mirror is quite appropriate in this circumstance.

But when considering 'vendor push' technology, beware that there may be a tendency to focus on technology which could be considered as 'toys which are fun to play with' rather than new technology which will really assist the business. These new and interesting forms of technology are often promoted as disruptive technology when they are really only just 'distractive technology'. That is, it may be just a solution (a really great application or idea) looking for a problem to solve without there being one available.

Some examples of Distractive Technology might include:

- Google Earth mash-ups – while Google Earth is definitely disruptive in concept and content, the extension of this into themed 'mash-ups' for basic mapping purposes has attracted considerable public attention as new and exciting. This is obviously very useful, but is this just a different alternative to presenting data normally associated with a corporate GIS environment on every desktop, albeit in a public environment?
- Location-based technologies – an interesting concept using GPS and other technologies to locate users, often mobile phone users, so that they can receive information (often advertisements) that is relevant to their location, e.g. a product which may be provided in a retail store that the user is about to walk past. It is an example of 'a solution looking for a problem to solve' and although it will probably be pervasive in the long term, particularly if made to work inside buildings, it still has some way to go.
- Telecommuting – for several decades technology vendors have been promoting the ability of staff to work from home and to telecommute to the office. This has been espoused as the nirvana of the modern workplace, but has yet to occur in any real manner – perhaps the telecommuting concept omitted to consider the need for staff to have human interaction with their fellow workers?
- Paperless Office – as one industry pundit mused several years ago, the 'paperless office will probably occur in about the same timeframe that the paperless toilet occurs', that is, – not in my lifetime. While this may happen eventually, it has not and probably will not happen because human nature and current work processes continue to require the need for paper, except in some specialised business environments such as insurance and banking.

Therefore, in developing a strategy one needs to remain focused to ensure that any disruptive technologies potentially being considered are actually useful and will assist in meeting the needs of the business, or whether it is just distractive technology that is otherwise clouding the business purpose.

4 Developing the Business Focus

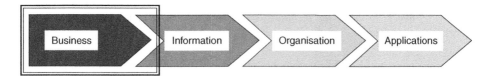

The development of any technology-based strategy should focus on the operational, business and statutory needs of the organisation and its individual business units, rather than that of the technology. This is necessary to ensure that the focus and resultant strategy (along with any subsequent technology) are aligned with business needs and not focused on a technology-based 'wish-list' of needs.

The concept of defining a small number of factors which are critical to the success of the business (i.e. Critical Success Factors or CSFs) allows managers and senior personnel to better undertake their roles by concentrating on those business issues critical to the performance of their role.

The role of CSFs in Information Systems analysis can assist principally in three areas:

- the definition of the business priorities and management information needed to support decision making within the business process;
- ensuring that the operational processing functions directly support the goals of the business process; and
- allowing the analyst to determine whether technology can be most effectively employed to support and enhance the effectiveness of the business process.

As such, CSFs allow analysts to ensure that the objectives of any technology are focused on ensuring that business aims are met, rather than to focus on technology items.

A review of an organisation's Corporate Plan, Annual Report and other corporate documents can often provide a useful first-pass outline of the high-level business direction of the organisation. However, these documents should be validated by interviews with senior management to ensure that they do actually represent the goals and objectives of the business.

One of the first steps in this process is to review the Vision statement, Mission statement and Key Result Areas (KRAs) of the organisation. All well-managed organisations will (should) have these documents and should continue to refine them as part of their ongoing annual business planning cycle. These documents also help to focus the organisation on the business plan as well as to focus the business plan on the business climate in which the organisation operates. Therefore, Vision and Mission statements (and the KRAs that support them) should clearly enunciate the focus of the organisation while enthusing staff and customers to work towards that focus.

Achieving Business Success with GIS   Bruce Douglas
© 2008 John Wiley & Sons, Ltd   ISBN: 978-0-470-72724-9

For example, the following Vision and Mission statements from a major organisation are reasonably typical of those that one would encounter in a number of organisations:

Vision: *To be recognised as a dynamic world-class organisation that is reliable, dependable and customer-focused, serving as a catalyst for regional economic development.*

Mission: *To manage and develop the organisation in a manner that facilitates existing customers to maximise their business opportunities and encourages new customers to utilise the facilities whilst earning a commercial rate of return for shareholders.*

While these Vision and Mission statements seem good (on first read), they could belong to any company. These statements convey only motherhoods which have very little meaning to the average staffer or customer. For example, these statements could apply to a food retailer, a government transport department, an electricity utility or a packaging company – all of these industries could 'lay claim' that these Vision and Mission statements represent their business.

And because many Vision and Mission statements are as vague as the examples used above, they are not all that relevant, which is why it is essential that the Vision and Mission be validated by interviews with senior management – sometimes the results of these interviews can outline directions which may be quite different to that expressed in the actual Vision and Mission statements.

Ideally, Vision and Mission statements should consist of words which convey the focus and direction of the organisation and use words that have an everyday meaning to those who deal with that organisation.

Once the correct Vision/Mission of the organisation has been understood, further drilling down should reveal how the business plan will meet the Vision/Mission through the development of KRAs and Goals to provide a focus to the organisation. This information is critical to the development of a technology-based strategy.

An example set of KRAs and Goals could be:

Key Result Area 1 – Trade and Industry (example)

1. *To increase and diversify trade through the facility to maximise financial and economic return.*
2. *To maximise financial and economic return to shareholders.*

Key Result Area 2 – Infrastructure, Services and Management (example)

3. *To provide reliable, appropriate, timely and competitive infrastructure and services to meet current and future trade and industry requirements.*
4. *To ensure all risks associated with the agency's activities and operations are managed in accordance with legislative and policy requirements to reduce exposure of the agency and stakeholders.*

Key Result Area 3 – Community and People (example)

5. *To be a responsible corporate citizen that is responsive to the needs and expectations of the community.*
6. *To have highly productive, enthusiastic, motivated and appropriately skilled employees.*
7. *To promote and maintain an integrated and coordinated regional community.*

Key Result Area 4 – Planning and Environment (example)

> 8. *To adopt whole of region planning and management for the sustainable growth of the facilities and the region.*
> 9. *To value the environment and amenity of the facilities and its surrounds.*

These KRAs and Goals should then be supported by the organisational structure/groupings responsible for delivering these KRAs and goals. The organisation may have a number of organisational teams and/or business groups, such as:

- Operational Services
- Corporate Governance
- Finance
- Engineering
- Planning and Environment

While each of these organisational teams (business groups) will, of course, be focused on their KRAs, it is important that each is mindful of the overall requirements of the organisation and understands how the organisation works.

Therefore, as part of the process to understand the objectives, issues and capabilities of each of the organisation's business units, a number of interviews and/or workshops should be held with key internal stakeholders to explore and focus on those requirements needed to meet the primary business drivers of the organisation, that is, the factors required for success.

These interviews should start at a reasonably senior level in order to obtain the correct perspective, to understand the broad business and direction of the organisation and to understand the key issues that impact (sometimes adversely) on the delivery of the stated business needs.

Hint

It is always best to have an external person undertake these interviews because this person will need to interview a number of senor managers, and:

- very few bosses will tolerate being questioned in some detail about their visions, plans, aspirations, fears, etc. by an employee;
- most interviewees will tolerate an outsider asking 'seemingly dumb' questions about the organisation and its business and will explain 'reasons why', but an internal person will most likely be told that he/she should already know that;
- internal interviewers will have preconceived ideas about the organisation and what it should be doing – some right, some wrong;
- internal interviewers would most likely not dare ask probing questions of a very senior manager as it would be seen as a 'career-limiting move', i.e. 'I could get sacked if I asked him/her about that'.

A key part of these interviews is to ask questions that seek confirmation (in the managers own words) about the two, three or four issues that are really important to their own success and that of the organisation, i.e. the CSFs. When this is confirmed across a number of interviewees, one can then start to understand how the CSFs reflect the actual business drivers.

Note that all interviews at this stage should focus on understanding the business of the organisation, the nuances of how the structure and business units work, the politics of the organisation, the major projects being undertaken, the competitive pressures, threats and risks, etc. In all of the meetings up to this stage, there should not be any discussion about GIS or technology per se, unless it is instrumental to the subject of delivering business benefits.

As these meetings progress, it will become apparent how the IT/GIS can (should) support their part of the business and how data (or the lack thereof) can impact meeting these goals. During this process the basic business drivers will become apparent. It will also become apparent that there is a chain of command in how the business drivers (CSFs) are fed top-down and can be impacted from bottom-up. All of these drivers need to be aligned in terms of tasks and targets. That is, just as an organisation has a hierarchy, so too do the CSFs.

The following diagram shows a typical organisational structure. A number of Business Units form part of particular Sections, which in turn form part of Departments, which go to make up the whole organisation.

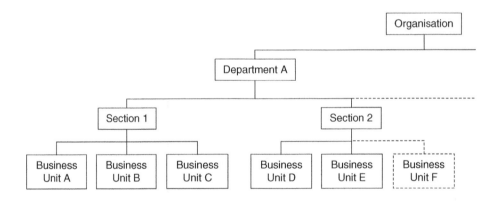

In this example, Business Units A, B and C have a sole purpose to meet the business objectives for Section 1. Therefore the CSFs of Business Units A, B and C must collectively form the CSFs for Section 1, which must form part of the CSFs of Department A, and so on.

However, the CSFs or business drivers are set at the top of the organisation and are typically segmented along organisational lines to create the CSFs for each Department, Section and Business Unit. This can also be extrapolated to a personal level whereby the Manager of a Business Unit should have personal CSFs which are aligned to those of the Business Unit which he/she manages. In this way, all individual and business unit CSFs should 'roll-up' to form part of the corporate CSFs.

In order to ensure that all issues are canvassed when trying to collect the business drivers or CSFs, it is important to look at the organisation without pre-conceived ideas and to check off the basics. Part of this process is to make sure that there are no obvious business drivers that are so obvious that they do not get discussed.

Case example

For example, several years ago I was undertaking a large project for a State-based Police, Fire, Ambulance and Emergency Services consortium. As obvious as it was, when I was interviewing a number of senior fire managers, none indicated that a primary CSF was 'to put out fires' – all had taken that as read and were focusing on issues such as resource planning and analysis. To them, 'putting out fires' was so obvious that it almost was not discussed.

In summary therefore, it is critical that the business analyst makes sure that the CSFs are correct before anything else proceeds. The CSFs will drive all future strategy development, particularly to correlate against the technical requirements (when developed) so that a focus for the GIS Strategy can be provided to meet the needs of the business.

Note that it is important to ensure that the CSFs are correctly identified:

- for the organisation as a whole;
- for each individual business unit; and, if required
- for each individual business unit manager.

In the example discussed above, the CSFs were logically based on the above KRAs and ratified during discussions with stakeholders and senior managers, summarised as:

1. To increase business operations and maximise financial and economic return.
2. To provide the appropriate level of infrastructure to support the business operations and outcomes.
3. To provide reliable, timely, accurate and integrated data/information to support the business, particularly the infrastructure and services requirements.
4. To minimise operational risk and exposure by ensuring that all data provided is correct, up-to-date and available.
5. To have trained and efficient staff using appropriate tools.
6. To provide better environmental outcomes for sustainable growth.
7. To be able to respond to community needs in a timely manner.

These CSFs will then be used for later correlation against the functional requirements (when completed) in order to ensure that each functional/technical requirement does actually support one or more CSFs. If a functional/technical requirement does not support a CSF, then that requirement would either have a very low priority or be deleted. Of course, each of these seven overall CSFs can be broken down into next-level CSFs as required, depending on the level of detail required.

5 Developing the Data/Information Focus

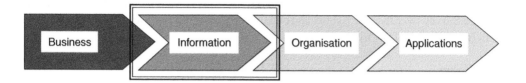

Geographic Information Systems are, as the name implies, information systems based on geographic data. They generally do not contain transactional data such as one would find in an accounting system where data is used for the current month, summarised and then archived. GIS data is (almost) never culled – it is always used, built-on, analysed and built-on some more.

Over time, and if properly implemented and managed, the value of a GIS to an organisation is that of the ever-growing data asset, rather than the value of the hardware or software. The GIS software is transitionary, but the data remains and grows.

Fact

For the last couple of years, the GIS/Spatial Survey has found that 40% of the GISs in the Australian/New Zealand Industry had been replaced in the previous (rolling) 3 years. This indicates that there is a high level of 'GIS churn' in the market and that it is continuing.

This high turnover of GISs highlights the importance of ensuring that the spatial data is portable from one GIS to the next, rather than focusing on the next system to be acquired.

As such, the data asset of a GIS in an organisation is cumulative – the sum total of all the preceding data, hence the expression that GIS is 'data-centric' or 'data-hungry'. Therefore, because the data will become a substantial asset over time, it is important to ensure that a correct focus is developed for the spatial data/information.

In addition, it is important to put measures in place which enhance the data and guard against the degradation of the data as the GIS software changes, particularly as the GIS database will (most likely) include data from other organisations, perhaps to a lower standard.

Achieving Business Success with GIS   Bruce Douglas
© 2008 John Wiley & Sons, Ltd   ISBN: 978-0-470-72724-9

5.1   INTRODUCTION

Spatial data has the potential to substantially impact (positively or negatively) the implementation of GIS, and because most GIS implementations are corporate in nature, GIS data is most often considered in the context of a 'corporate database'.

While the spatial data used in a GIS environment varies from organisation to organisation, the following is a list of datasets which have been reviewed by the GIS/Spatial Best Practice Survey over the last 7 years and consistently found to represent the needs of most users (not in order of priority):

1. Cadastre	20. Land Use Strategic Plans
2. Land Title/Ownership/Lease	21. Demographic Data
3. Geodetic	22. Forests & National Parks
4. Native Title Claims	23. Parks & Recreation Areas
5. Topographic	24. Cultural & Heritage
6. Digital Terrain/Elevation Models	25. Flora & Fauna
7. Aerial Photography	26. Vegetation Boundaries
8. Satellite Imagery	27. Significant Trees/Vegetation
9. Roads	28. Hazardous & Contaminated Sites
10. Pathways/Cycleways	29. Soils & Geology
11. Bridges	30. Utility Data – in-ground
12. Street Lights & Furniture	31. Utility Data – above ground
13. Towns & Places of Interest	32. Utility Data – plant & equipment
14. Bus/Ferry/Tram/Train Routes	33. Assets, Buildings & Facilities
15. Street Addresses	34. Easements
16. Admin Boundaries (electoral, shire)	35. Mining Areas
17. Postcode Areas	36. Maritime/Oceanographic Data
18. Suburbs	37. Atmospheric/Meteorological Data
19. Land Use Zoning Plans	

Very few organisations would use the majority of these datasets at any one time, however most do use a number of these datasets, depending on the focus of the organisation. The datasets which are endemic to almost all GISs, of course include such data as the property framework, topography, roads, towns and administrative boundaries (e.g. suburbs and electoral boundaries, etc.).

5.2   METADATA

When data is loaded into a GIS environment, it often loses the 'context' from where it was derived, and may conflict with other data collected at a different time by different processes or to differing standards of accuracy. In order to understand the linage of each data item, it is important that information about the data (metadata) is entered into the GIS so that, over time, knowledge about each piece of data is not lost and can be used to ensure that the data is 'fit for the purpose' to which it is planned to be used.

Generally metadata is required at the level of the individual data item rather than the whole dataset. If metadata is not collected, the quality, completeness and usefulness of that data item is often unknown, thereby making some of this data useless or, at best, of lesser value than it should be. Metadata, properly applied, should be a commonsense short description containing useful information such as:

- who captured the data (e.g. Bill Smith, Planning and Environmental Section);
- the reason why the data was captured (e.g. for environmental reporting);
- the date of capture (e.g. March 2003);
- how the data was captured (e.g. by hand-held GPS, by digitising or from log sheets);
- the source material (e.g. 1997 civil drawings, field capture, 2002 aerial photos, etc.);
- the precision of capture (e.g. to a precision of 20 mm, or ± 5 m, etc);
- the locational area of the data (e.g. Central Station, NE Western Port Bay, etc.);
- the fitness for purpose (e.g. captured to provide a rough outline for planning);
- the completeness (e.g. all available data was captured or only a quarter was captured); and
- the coverage of the data (e.g. all of Lease xxx, Pipeline data only).

Metadata should be stored with the data and be readily available to the user of the data for him/her to make a decision, based upon the metadata, about the appropriateness of that data to meet the task at hand. This requires that metadata should be input by the person capturing, collecting or converting the data. To ensure that metadata is captured, it is essential that the metadata is easy to input and amend. Therefore, it is often necessary to provide a mechanism for metadata to be easily input and for the metadata to be stored with the data.

5.3   DATA/SYSTEM ARCHITECTURES

While data must meet the needs of the business, one of the major issues to be considered when discussing data in relation to developing a GIS Strategy is that of the data architecture.

Because GIS is data-centric, the GIS software must be able to work closely with the GIS data for the system to be efficient. In a number of cases, the software and the data are so inextricably bound together that it is difficult to separate the architecture of the data from the architecture of the system. This may sound confusing, but the best way of explaining this is to consider the following typical types of GIS implementations (software and data):

1. *Standalone GIS – the single database (and simplest configuration)*
   A GIS application is installed on a single PC and the data is loaded on the same PC. The GIS does not exchange data with any other system, and can be considered as being a 'closed loop' system. The user loads data, edits data and produces output. All data is stored on the disk of the PC and the integrity of the data is as good as the diligence of the operator. In this environment, any performance issues are the result of the capability of the single PC and the size of the database, and can usually be solved by upgrading the PC.
       In this scenario, data is typically stored as graphic files in the proprietary format of the GIS being used. Being a standalone system, the GIS data is typically only

accessed by a single person. However, the data can be 'served' to the organisation by uploading this data periodically to a web server for deployment across the organisation's intranet.

2. *Networked GIS – small co-located configuration*

The next step up from the Standalone GIS is usually when the GIS is installed on several PCs and there is a need for each of the users to access the same data. The typical scenario used in these circumstances is that one of the PCs (or a server) is designated as containing the 'master' database and the remaining users 'read' the data from this master database and write back any changes to the data. Typically the GIS software is stored and launched from each PC.

If the PCs are located near each other (i.e. within a few metres), the speed of the several PCs and the small network are usually not adversely impacted and if there are any performance problems these can be solved easily by upgrading the PCs, server or the network.

However, because a number of users may be accessing the same database or graphics files at the same time, there is the possibility of two or more users editing the same data item at the same time. This would result in that data item (when saved by each user) having his/her changes overwritten by the user who saved last. These types of problems are usually avoided by each user working on a different theme of the database (e.g. environmental data or civil data) or on a different geographic part of the database when editing data at the same time. When all else fails, the users talk amongst themselves to avoid the situation.

In this scenario, data is typically stored as graphic files in the proprietary format of the GIS being used. In this configuration a small number of users access the GIS data, but again the data can be 'served' to the organisation by uploading this data periodically to a web server for deployment across the organisation's intranet.

3. *Networked GIS – co-located configuration and transaction management*

When GIS environments become too big or too complex for data management by the above method (i.e. talking amongst themselves), the next level of data management sophistication is often to implement a transaction management capability.

In a number of business environments there is a need to undertake transactions on the GIS database in much the same manner as a cash register transaction or the processing of an emergency call for police assistance. The transaction management capabilities used in several GISs rely on a 'job' being opened, actions being undertaken on the data, the job edits being 'posted' to the database and then the job being closed. A consequence of this philosophy is that maintaining proper chronology is extremely important. All data must be processed with chronology in mind, all processes (transactions) must have a start and an end and, although long transactions are possible, it is expected that there are not too many long transactions and that they do not last too long.

If a conflict occurs when two or more users try to post edits on the same feature, the transaction management software usually has the ability to reconcile these conflicts. The two main transaction management philosophies are:

• GIS systems which lock the data records to the first user so that others have only read access to the data, that is, second (and third) users cannot edit the data while the first user has the data;

- GIS systems which prefer to allow everyone access to the data and then to resolve the conflict at the end of the process, rather than locking the second or third users out of the process altogether.

Regardless of whether one agrees or disagrees with these philosophies, the user and system manager must work within the context of the philosophy of the GIS product being used.

In this scenario, data is typically stored as graphic files or as a graphic database in the proprietary format of the GIS being used, but is tightly managed by the transaction management capability of the GIS. Again, all users have access to the GIS data, and the data can be 'served' to the organisation by uploading this data periodically to a web server for deployment across the organisation's intranet.

4. *Networked GIS – distributed configuration*
When the GIS is required to be distributed between several offices in the one city or between offices in several cities, the data management strategy can suddenly acquire a substantially higher degree of difficulty.
The following scenarios often present themselves in a distributed environment:

a. If the network is fast, the distances between the offices are not great (i.e. less than 500 m) and the network traffic is not too great, then the scenarios outlined in (2) or (3) above can be used. The data can continue to be stored as graphic files or as a graphic database in the proprietary format of the GIS being used and the data can continue to be stored at the central server and accessed at the remote workstations.
b. However, if the network is slow, the network traffic is too high or the distances between the offices are large, then the performance of the GIS may be such that it becomes unworkable. In this eventuality, the following options can be considered:

- Upgrade the server and network and implement a product such as Citrix to enable the remote workstations to continue to access the central database in real-time (this is similar to the old mainframe concepts of the 1970s and 1980s, but with upgraded technology).

  In this scenario the data can continue to be stored as graphic files or as a graphic database in the proprietary format of the GIS being used and the data can continue to be stored at the central server and accessed by the remote workstations in real-time.
- Use a web-based client such that the user can use the GIS and edit data via a web connection. While this is a useful option if the data is edited rarely and if the web connection is fast, this option may be unworkable for a heavy workload and/or if the web connection is slow.

  In this scenario the data can continue to be stored as graphic files or as a graphic database in the proprietary format of the GIS being used and the data can continue to be stored at the central server and accessed by the web clients in real-time.
- Consider a database strategy based on the replication (copying) of the database at periodic times (usually nightly or weekly), with the database changes being managed by a real-time 'long-transaction' management process (refer above).

  In this scenario, the data can continue to be stored as graphic files or as a graphic database in the proprietary format of the GIS being used but the data is stored at each remote location and replicated to the central server at periodic intervals for reconciliation and re-distribution of the 'master copy' to the remote sites. Because data is copied between servers, the versioning and management of this data becomes very important.

In all of these scenarios, the data is typically stored as graphic files or as a graphic database and tightly managed. Again, all users have access to the GIS data, and the data can be 'served' to the organisation by uploading this data periodically to a web server for deployment across the organisation's intranet.

5. *Networked GIS – DBMS option*

   An emerging trend in the SI industry is to store spatial data in a spatially enabled relational database such as Oracle Spatial, or in conjunction with 'spatial engine' middleware products such as ESRI's SDE or MapInfo's SpatialWare. This often greatly enhances the methods and approaches that can be used to deliver end user functionality and maintain the underlying data, while facilitating the adoption of future advances in spatial technology.

   The following RDBMS technologies are often used for these scenarios:

   a. Oracle RDBMS – while offering some basic spatial functionality within the standard product, further functionality can be acquired in the Enterprise edition that extends spatial capability as well as interoperating with several 'spatial engines' provided by major GIS vendors;

   b. IBM RDBMS – the merging of the spatial functionality migrated from Informix spatial datablades with that provided by DB2 can provide some functionality comparable with Oracle Spatial, albeit without the extensive user base of Oracle Spatial; and

   c. SQL Server RDBMS – while not spatially enabled, SQL Server does interoperate with the spatial engines provided by the major GIS vendors.

The advantages of this approach are:

- in distributed environments, the 'bandwidth' required for transferring spatial data stored as textual objects in an RDBMS is quite small (and therefore may be used to avoid a strategy requiring database replication);
- if replication is still required, RDBMS platforms often have a better replication strategy than a replication strategy based on a GIS vendor using proprietary formants (alternatively, another form of replication is simply to replicate the whole database, folders or servers as binaries, without the replication software being aware of the replication); and
- the data stored in the RDBMS becomes 'independent' of the particular GIS proprietary vendor format, thus becoming an effective neutral format to exchange data with other GISs which may be using different software with different proprietary vendor graphic file formats.

The disadvantages of this approach are:

- managing an RDBMS environment, such as Oracle Spatial, is a complex and highly skilled task requiring a higher level of trained staff, often at a cost which could be substantially higher than other technology environments, in addition to the additional cost of the RDBMS;

- some of the spatial data structures (e.g. for dynamic segmentation) inherent in some GIS products may not be able to be stored as a spatial object in an RDBMS environment, resulting in that data structure not being able to be used for this strategy; and
- not all software platforms support storing data in an external RDBMS.

For this option, all data is stored in an Relational Database in a format which is independent of the proprietary graphic file format or graphic database format of the GIS being used, although it is still in a format which is proprietary to the RDBMS being used, albeit this is simply a table structure which is easily transferable. Again the data is in a structure which is tightly managed by transaction management and/or replication and, although all users have access to the GIS data, the data can be 'served' to the organisation by uploading this data periodically to a web server for deployment across the organisation's intranet.

Of course, there are many variations to each of these five broad scenarios, a number of which are dependent on the particular GIS vendor products being used. However, the point being made here is not about the technology sophistication per se, but that as each level of complexity is applied to a GIS the impact on the data and the method of storing and accessing that data can be substantial.

In addition to these discussions about data architecture scenarios, one must also realise that not all spatial data is volatile. Most spatial data is static, i.e. it may not change, and the data which does change may be quite small, e.g. less than 5% of the total database size. Therefore, whichever database strategy is selected from those discussed above, if less than 5% of the data is changing, a viable alternative may be to copy the 95% of the data (which is static) to each location and then partition the volatile 5% and consider a database strategy which addresses only that data which is volatile.

In summary, the data architecture can be an issue which will substantially impact a GIS Strategy and it is therefore important to get it right so that the data architecture is appropriate for the initial GIS implementation as well as providing the appropriate foundation for a final implementation which may be several years away (and after a suitable data migration). Again, because GIS is 'data-centric' then so must be the GIS Strategy.

5.4 DEFINING THE 'DATA GAP'

Because data is a critical component in implementing GIS, it is important to consider:

- the current data store: e.g. GIS, CAD, textual, etc.; and
- the data required to meet the needs of the GIS.

That is, in the context of a GIS Strategy, particularly one requiring a cost/benefit, it is 'the data gap' that is crucial to be defined and costed (including the cost to convert existing data such as CAD data from one form to another). This data gap is therefore:

(The data required to support the forecast business outcomes of the GIS Strategy)
minus
(The data currently available)

REQUIRED DATA TO SUPPORT THE GIS STRATEGY

At a conceptual level, and in the first instance, I find that it is often useful to summarise the major datasets which are required to support the GIS Strategy along the lines of the following short descriptions:

1. *Dataset: Lease and Cadastral Boundaries*

Priority:	Very high.
Purpose:	To depict the lease and cadastral boundaries of all relevant land parcels in (and surrounding) the area of interest.
Comment:	The lease boundaries have been captured in AutoCAD and the surrounding cadastre is available from the local Councils. This includes Lot parcel polygons, street boundaries, attribute information (Lot on Plan number) and a range of administrative boundaries (e.g. Suburb, LGA, Electoral Boundaries, etc.).
Coverage:	The coverage required for this dataset is the facility area and surrounding area.
Custodian:	In-house for Lease polygons, Council for surrounding parcels and State Government for other cadastre.

2. *Dataset: Major Buildings/Plant Infrastructure*

Priority:	Very high.
Purpose:	To show the location of all major buildings and plant infrastructure in the facility area and surrounding area.
Comment:	While most of this data is available either from the drawing archive or from CAD databases, it needs to be a concise set of data about the facility which can be used for planning, operations, maintenance and environmental management, therefore it is suggested that it should be collected from the primary source – either from ground surveying or from aerial photography.
Coverage:	The coverage required for this dataset is the facility area and surrounding area.
Custodian:	In-house and some Council.

Note that the summarisation shown for the two datasets above should be undertaken for *all* datasets, before any further detailed assessment is undertaken.

Following an initial assessment of the data required, the data available and the consequent gap, detailed planning can commence to determine how the data gap can be remedied while still ensuring that the data needs can be correlated against business needs.

The most common method to undertake this process is to form a committee of stakeholders to consider the major issues such as business processes, corporate data, etc. and how these might support the business, i.e.:

- What is corporate data?
- Who uses it?
- What are the business reasons for capturing corporate data?
- How much duplication of data capture is occurring?

As part of this process, the conceptual summaries for each dataset are workshopped with stakeholders so that further detailed specification can be applied (e.g. detailed specification of the geographic coverage, data content, data accuracy, data currency, etc.) to such a level of detail that the capture/conversion of the required data can be adequately costed.

Case example

As part of the implementation of a GIS Strategy across a large multi-function government agency with diverse requirements, I was facilitating a Data Consultative Committee process to get the stakeholders to discuss their data and the business reasons behind the need to collect/maintain it.

However, the Departmental Managers did not want to get involved, because this 'was only about data'. Unfortunately they failed to understand that this was 'about the information necessary to drive their business' and if they did not have the right information to run their business, then the GIS would not be able to focus on their business needs, and they wouldn't get the benefits promised.

So involvement on this committee was delegated down to their subordinates, who also did not understand (and were too busy) so they delegated it down until only GIS Officers attended the meetings (and only because they did not have anyone under them to delegate it to). In good faith, and with a total lack of direction from their management, these GIS Officers were making monumental decisions about which data to capture, what attributes were important and to what level of accuracy the data needed to be acquired.

Needless to say, this process went nowhere because it lacked management input and support. The Parks people were collecting street tree data for maintenance purposes while the Stormwater people were collecting tree locations for pipe interference checks and the Planning people were collecting tree locations for the development approval process. And they were all doing this at much the same time, in triplicate, with different standards of capture.

Such a Data Consultative Committee can ensure that the correct business focus is applied to decisions about data, particularly to ensure that those data decisions align with the business requirements and outcomes. But beware that a committee of stakeholders often has the downside that there is a need to manage the risk that the committee will be all talk and no action.

EXISTING DATA HOLDINGS

The current data holdings of most organisations can be categorised as follows.

Existing GIS Data

An organisation with an existing GIS will have some data which should be able to be migrated into a new environment. This should be catalogued on a data-type by data-type basis with indicators on accuracy, precision, currency, fitness for purpose, metadata, etc. in order to determine the usefulness of bringing the existing GIS data forward into the new environment.

Existing CAD Data/Drawings

Most organisations who implement GIS have a considerable amount of data stored in CAD drawings, often as AutoCAD or Microstation files of civil, electrical, mechanical and structural drawings. Generally only those drawings with a civil component would be useful to migrate to a GIS environment and only that part of the drawings that were geographically 'reasonable'. This data would need to be converted to the GIS coordinate system being used after being imported from the .dxf (or .dwg) AutoCAD file or .dgn Microstation file or other CAD data formats.

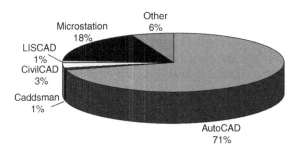

Results of the GIS Survey show that the distribution of CAD systems used in GIS environments is as shown in the previous chart. AutoCAD was reported as being used by almost three-quarters of the GIS industry, with MicroStation reported as being the other dominant CAD product used in conjunction with GIS.

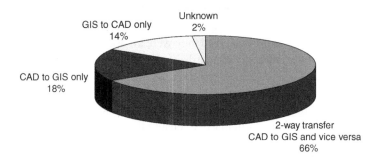

Almost all GIS contributors to the GIS/Spatial Surveys reported that they shared data between their GIS and CAD, with almost two-thirds of this data-sharing being two-way. The mode of data-sharing across all industries was as shown in the previous chart.

The method of exchanging data between the GIS and CAD environments was reported as shown in the following chart. Commercially available translation products were the most prolific method, however the percentage of those having achieved data compatibility between the two environments is 35%, thereby negating the need for data translation. The level of use of Custom Translators has remained reasonably static over the last several years.

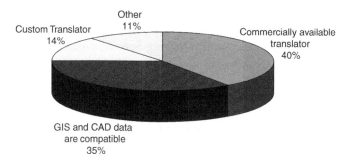

Existing Drawing Images

Typically there may be a large number of drawings in the organisation which may have been scanned and would be available as JPEG, TIFF or GIF images. These could include images of civil, electrical, architectural and planning drawings, either as a by-product of the CAD process or as an image of a drawing which exists in hard-copy only (and has been imaged).

This image base often represents a wealth of data which the organisation can use to augment the GIS data as required. However, because these images are often not geometrically compatible with the GIS, it is common that these drawing images remain as images that are opened only in a separate window for the operator to reconcile (mentally and manually) with the data in the GIS on an as-required basis.

Existing Aerial/Satellite Imagery

A number of organisations may also have aerial photography and satellite imagery which could be available for use in the GIS. Again, for these images to be useful, they would need to be ortho-rectified and geographically positioned/oriented/scaled so that they can underlay the data in the GIS. The currency and scale of these images should be reviewed to ensure that they are relevant and fit for the purpose to which they are intended to be used.

Existing As-constructed Data

Over the years we have found that as-constructed data (usually in the form of AutoCAD drawings) forms the dataset which is most frequently not gathered by organisations or not completed properly by contractors, even when it is gathered and submitted. Often as-constructed data is little more than as-designed drawings notated 'everything was placed as per the design' – often quite at variance with reality. As-constructed data represents the record of works done by contractors on work-sites and is therefore a valuable record of the location and condition of assets and services but only if it is correct. While it is also a legal record of work done, it is also an excellent source of data for GIS.

However, because there is often a work/responsibility dichotomy between GIS staff and staff who receive the as-constructed data in large organisations, as-constructed data is either not collected or is collected in a form which is not useful to be entered into a GIS. In order to facilitate the easy entry of data from contractors, including designs, developers' plans, etc., contractors should be encouraged/mandated to provide all information in an electronic form that can be imported into the GIS quickly and easily while meeting the standards required for the GIS.

In order to facilitate this, an Electronic Drawing Specification often needs to be developed which can be used as the basis for specification of a data import tool as well as for use by contractors to ensure that data from these sources is aligned with the methodology that the GIS uses for storing and categorising data.

To facilitate the take-up of this specification by contractors, it is often useful for the organisation to develop a set of CAD routines for use by the contractors in conjunction with their CAD software so that a CAD export file is created with the correct symbology/layering/attribution, etc. to meet the requirements outlined in the Electronic Drawing Specification.

Other Existing Data

Other data, such as asset data, planning data, network data, environment data, hydrographic data, etc., has the potential to be a valuable resource for the GIS – such data needs to be sorted, checked for compatibility, currency, relevance, fitness for purpose, etc. before bringing this data forward for use in the GIS. Industry experience is that much of this type of data exists in a range of Excel spreadsheets (or Access databases), often duplicated with other data and often requiring some considerable amount of effort to decipher before a final decision can be made as to whether this data is useful or not.

In order to address these issues, some organisations are undertaking data quality improvement projects (as time and budgets permit) to clean up old legacy data rather than re-capture the data from the source. While this is often a difficult task, the cost of clean-up can be substantially less than the cost of re-capture, particularly when clean data can substantially improve the quality of decisions resulting from the GIS.

Existing Data Summary

In summary, while there is a general need to catalogue the existing amounts and types of data used within the organisation, a high-level overview is often useful to further discussion about whether existing data is useful and likely to be carried forward into a GIS environment.

When the quantum of existing data which can be brought forward into the GIS environment is known, then consideration of the gap between the existing and proposed environment can be developed.

DATA GAP SUMMARY

In summary therefore, it is the 'data gap' which needs to be addressed in the development of a GIS Strategy, since there will always be existing data and that data is not always enough. That is, the data gap is the amount of data required to support the GIS strategy (and its planned outcomes) minus the data currently available.

It is often the case that a lot of the information required for a corporate GIS is either not available or is so dated that in a number of cases there would not be a lot to be gained by undertaking a detailed data cataloguing process, except for that which cannot be gained from other sources (particularly relevant for converting existing drawings of underground assets). Note that any data conversion process should also include an 'archive and forget' function for data which is no longer of use.

5.5 GIS DATA STANDARDS AND RELATED ISSUES

No discussion on GIS data would be complete without a short discussion on the standards which relate to GIS data. Data standards in the SI industry can be traced back to a number of earlier initiatives of the CAD industry in the 1970s and 1980s, such as the International Graphics Exchange Standard (IGES).

In the 1980s and early 1990s, the early GIS data transfers were mostly based on localised formats for exchanging digital (often topographic and property) data. In Australia, the standard AS2482 (Interchange of Feature Coded Digital Data) was based on transferring contour data for national mapping and military purposes and as such was only concerned with the types of data that one might find on a topographic map sheet at a scale of 1:100 000 or similar.

Unfortunately most jurisdictions that developed data exchange formats all did so in a different manner, with the end result that one needed to use a different standard for every organisation that one provided data to or received data from.

In the early 1990s, the Spatial Data Transfer Standard (SDTS) was developed by the US Geological Survey (USGS) and US Defense. Because SDTS was an all-encompassing standard designed to meet the needs of all GIS users from all industries, it was very lengthy, quite difficult to understand and very difficult to use. Consequently very few organisations actually used SDTS.

Case example

SDTS was adopted and Australianised by Standards Australia/New Zealand in the mid-1990s and at that time the Australian and New Zealand Land Information Council (ANZLIC) put considerable effort into the promulgation of SDTS through funding a 3 million dollar initiative to provide education, support and development of the standard.

To my knowledge, Geoscience Australia (Australian Federal Government) and the Western Australia Department of Land Information were the only organisations which produced data in SDTS format, with other States and Federal agencies largely ignoring this initiative, notwithstanding that most of the same State and Territory agencies, being represented on ANZLIC, had approved and jointly funded this promulgation initiative.

So while SDTS existed as a standard and was operational, almost all government agencies (at all levels) in most countries largely ignored it. Consequently most vendors did not put much effort into developing translators to provide data in SDTS format.

Therefore while SDTS was the (then) standard, without translators is was largely ineffectual. This resulted in almost all data exchange continuing to be undertaken in an ad hoc manner, usually using the 'lowest common dominator' format, in many cases using either

.dxf (AutoCAD's exchange format), ESRI's SHAPE file format or MapInfo's TAB file format.

In the mid to late 1990s, a number of organisations started to develop a range of translators to move data between the proprietary formats of each GIS that they came into contact with. One of the most successful universal translators developed was the Feature Manipulation Engine (FME) software by Safe Software – this provided the ability for organisations to move data from any GIS to any other GIS. FME continues to be the translator of choice in most GIS environments today.

However, substantial work continues to be undertaken by standards-based organisations in many countries on a number of issues relevant to the industry. Over time, standards have emerged on a range of different categories, such as symbology, topological structuring of data, metadata for imagery, marine metadata profile, bibliographical elements on maps, encoding, etc. In addition, a good deal of the work has been undertaken in the area of developing protocols adapted from the IT industry (e.g. eXtensible Markup Language or XML) for use by other industry sectors (e.g. Geographic Markup Language or GML).

In parallel with this, standards also continue to be developed in most disciplines based on representing and conveying information relevant to that discipline, e.g. standards for representing information (often spatial) in electrical engineering, mapping, civil design, etc., often resulting in that data being used in GIS environments. In fact, there are so many standards being developed that one sometimes has a choice about which particular standard to use, perhaps diminishing the value of standardisation.

The Open Geospatial Consortium (OGC) has also been active in the development of service-oriented specifications which may become used as standards in the GIS industry. Two examples of OGC specifications which are being adopted by vendors are quoted below:

- *Web Map Service (WMS) specifies how individual map servers describe and provide their map content of one or more maps from one or more map servers described in a portable, platform-independent format for storage in a repository or for transmission between clients. This includes information about the server(s) providing layer(s) in the overall map, the bounding box and map projection shared by all the maps, sufficient operational metadata for Client software to reproduce the map, and ancillary metadata used to annotate or describe the maps and their provenance for the benefit of human viewers.* [1]
- *Web Feature Service (WFS) defines interfaces for data access and manipulation operations on geographic features using HTTP as the distributed computing platform. A web user or service can then combine, use and manage geodata from different sources by invoking WFS operations on geographic features and elements.* [1]

While the Open Geospatial Consortium is not a standards organisation per se, it has representation from most of the vendor community and is working towards similar outcomes as most standards organisations. The take-up of WFS and WMS depends to a large extent on each GIS vendor developing connectors to that service, and, even though the consortium is largely comprised of vendors, these developments have yet to be completed for a number of GIS products.

---

[1] Open Geospatial Consortium Specification.

While good work is progressing on the use of WMS and WFS, some limitations have been encountered as the maturity of this technology increases, particularly when used in conjunction with different levels of maturity of different GIS products.

5.6 GIS DATA INTEROPERABILITY

The ability of different GISs to interoperate and to swap data has been an issue at the heart of standards development for some time. In a number of organisations, interoperability is key to managing risk, improving data integrity (re-use of data from other organisations rather than re-capturing it) and identifying opportunities (using data collected by other agencies).

But the lack of data interoperability between most organisations exists at a number of levels, not all of which are easily solved. For example, if one wanted to move data from a Water Utility using (say) an Intergraph GIS and from a Planning Agency using (say) a MapInfo GIS into the GIS of a Municipal Council using (say) ESRI, a number of problems emerge:

- the transfer of data in an Intergraph proprietary GIS file format and data in a MapInfo proprietary GIS file format to data in an ESRI proprietary GIS file format will be difficult;
- maintaining the topology (connectivity) between the data when it has been transferred (e.g. ensuring that individual road segments still connect as a continuous road) will be quite difficult and may require the need to re-connect data which may have become disconnected or lost in translation;
- maintaining the attributes (or connections to attributes) associated with graphic entities in each of the systems or in a related RDBMS (e.g. SQL Server) will be quite difficult;
- moving metadata from one system to the next (assuming that each GIS has metadata attached to the data items) will be quite difficult and if no metadata is available then a prudent user would attach at least the data source name (e.g. the Water Utility) as a metadata item;
- reconciling data captured at different accuracy resolutions will be very difficult, particularly when data from these agencies will probably be collected for quite different business purposes (e.g. reconciling very accurate utility locations with inaccurate or vague planning data);
- harmonising the naming conventions is often very difficult, that is, a footpath in one system may be called a trail in the next system and may be called a sidewalk in the third system;
- harmonising the same data themes which are stored on different layers / levels in each system will be difficult but not impossible; and
- having different graphics for the same symbol in the three different systems will cause a confusion that is best avoided.

While products such as FME can map the 'from/to' data paths, this has to be set up with some care and will change from organisation to organisation. In some cases this is exacerbated by different industries being required by law to show different data items in different ways on maps and plans – for example, the Electrical Engineering industry has a very strict set of standards representing how electrical symbology is presented.

Therefore, moving data from one system to another is a non-trivial and problematic exercise and one that is fraught with the need to often undertake a considerable amount of manual work to fix problems which might not have been forecast. It is not something that is done in a fast and ad hoc manner, particularly if quality data is required at the end of the process.

DATA INTEROPERABILITY IN EMERGENCY SERVICES

With the current political focus on terrorism-related activities, there is a very urgent need for the GIS in many levels of Government, Utilities and major industries to be able to exchange data with the GIS in almost all industries, particularly for law enforcement agencies and emergency services.

Case example

The September 11, 2001 attacks on the World Trade Center buildings in New York highlighted the data interoperability issue quite poignantly.

While the location of underground rail networks, power and gas utilities, trunk phone cables, underground car-parks, underground gasoline tanks, etc. was known by each utility, the correlation of these services was not known for a number of weeks and months after the attacks.

This was because each of these organisations stored their own data in their own IT and GIS systems, totally unrelated to the data from other agencies or utilities. If all of this information was available and in a common GIS form before the attack, it was estimated that relief work could have been substantially hastened.

As discussed above, data exchange is a very difficult and time-consuming task, therefore having data interoperability between a large number of different stakeholders in emergency situations is a very difficult task.

On initial consideration, it would appear that there are two ways in which data interoperability could be achieved, particularly to mitigate the risk of not having access to all requisite information to attend to an incident:

1. Either put all data into a single GIS and maintain that data in perpetuity 'just in case'; or
2. Develop the procedures and protocols to effect the seamless interchange of spatial data between the different systems as and when needed (i.e., a 'just in time' approach).

Option 1 would come at a high cost to mitigate an event that (hopefully) may never happen, but is a path that is being followed in many countries in response to terrorism-related concerns.

Option 2, based on data interoperability, seems to be the wisest course of action, so that when or if a major event does occur the procedures and protocols are in place to effect a rapid dump of data to a common system, if required. However, this option is the most difficult because it involves understanding different business processes and understanding the need that organisations have for storing data in specific forms for their business purposes.

Data interoperability in a major emergency situation may involve as many as 10–20 stakeholder organisations (several Councils, Police, Fire, Ambulance, Roads, Planning, Public Health, several Utilities, Phone companies, etc.). These issues can be considered as Technical and Organisational, as follows.

Some Data Interoperability Technical Issues

There are a very large number of technical issues to overcome when trying to amalgamate data from a large number of organisations – all of which have captured their data to different levels of completeness, different precision, different projections, different (data) structure and naming conventions, etc. for widely varying business uses.

As discussed earlier, the results of the GIS/Spatial Surveys have indicated that over 90% of spatial and aspatial data is stored in the proprietary format of the particular GIS product being used – therefore, when trying to amalgamate data from the GISs from a large number of organisations to a common GIS environment, typically the method of moving the data from one system to the other becomes via a lowest common dominator format with the consequence that a lot of the 'smarts' of the data (such as connectivity) often become lost in the translation process.

And because this process takes some considerable time and effort to effect (often weeks or months), it is not repeated frequently. As such, the data can become dated, with its usefulness thus diminished. Therefore, to ensure that the data does not get overly dated, it is important to put in place a mechanism to ensure that the data which does change in each of the (up to 20) organisations is provided to the common GIS on a regular basis.

All of this process, of course, takes considerable time and comes at a considerable cost to plan for an event that may never occur – and if it does occur, how will this data be used?

Some Data Interoperability Organisational Issues

On the assumption that there was one common GIS which held all the data from all other GISs for the subject locality (e.g. city) and that this data was correct and up-to-date all the time, would this really solve any problems? When major incidents do occur (e.g. a major rail bombing), the police or emergency Services do not attempt to re-route (or stop) trains or enter rail tunnels to disconnect power, gas and water, etc.

All major incidents have involvement from over a dozen relevant agencies, including the police and emergency services. In this scenario, any decision to re-route trains or disconnect power services would be undertaken by the rail or power company at their control centre.

So when this line of thought is followed, the question that soon becomes obvious is 'why do the police and emergency services need detailed data on the power lines when they do not have the expertise (or control) to be able to do anything with that data'. Indeed, one could argue that having access to detailed data would actually complicate and confuse issues and could slow down the response and recovery process, rather than assist it.

The existing processes to have the relevant rail / power experts on hand to coordinate their response through to their utilities control room (and GIS) has been used for a number of decades and has stood the test of time. Therefore, one could argue that emergency services do not need the detailed spatial data at all, but just the summarised locational data to understand the co-location relationships and the responsibility contact details for the services involved.

This then becomes less of a 'data interoperability' issue and more of a 'messaging interoperability' issue, so that the right person from the right utility is informed at the right time to interrupt the right service and that there is sufficient knowledge about the co-location of other utility services to ensure that there is no adverse impact from another utility service.

5.7   SUMMARY – DATA INTEROPERABILITY

Spatial data interoperability needs to be addressed, but not for many of the emergency services reasons discussed above, nor by building all spatial data into a single system – thus effectively avoiding the issue.

While data interoperability has been promoted by some in the industry as a Holy Grail, and one to which considerable resources should be dedicated, it could be argued that some of the current initiatives may be mis-aligned to the business and operational needs of the involved agencies, with the potential to divert considerable effort from that which does need to be addressed on the complex issues of spatial data interoperability.

True interoperability of data from disparate systems will (eventually) be helped by the work undertaken by the many standards agencies and organisations such as the OGC, but the practical implementation of such concepts will rest with the vendors, who will still be required to develop capability (connectors and translators) to / from their products to competitive products.

5.8   SUMMARY – THE DATA/INFORMATION FOCUS

In summary, it is important to understand the data environment of the proposed GIS, should the GIS Strategy be implemented. This data environment will most likely be focused on taking a corporate approach to data, and therefore addressing the organisational issues that this involves, and be based on addressing the 'data gap' by data capture, conversion or purchase.

The data architecture will be a critical component to developing this data focus. While the GIS may start small, it will almost certainly grow over time, and it is important to ensure that the data focus can grow with the GIS, perhaps transitioning the data architecture from a 'small co-located configuration' to a 'networked distributed configuration' over time.

Because the data focus is usually a significant cost component of the cost/benefit analysis of the GIS Strategy, it is important to ensure that all data items are specified in sufficient detail so that a practical and realistic cost can be derived.

The outcomes of this data focus are typically used as a component of the Functional Requirements Specification (FRS), usually an output of the Application and Technology considerations.

6 Developing the Organisational Focus

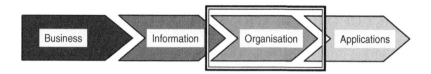

Any discussion on business drivers and the development of strategies to implement GISs should also include a section on organisational structure, since the structure of the organisation has the potential to severely constrain/impact the implementation of the technology-based strategy. This is particularly important for a GIS strategy which requires access to spatial data that is typically corporate in nature but held in diversified locations across an organisation.

Because GIS is 'data-centric', it needs a considerable amount of data to be loaded in order for it to be able to function and to provide information which is required to meet business objectives. This data is usually derived from a number of areas within an organisation and if the data is incorrect or poorly maintained, the decisions resulting from the use of this information will be less than satisfactory, and in some cases may be quite erroneous.

Therefore access to data which is corporate in nature, but held by individual business units, is critical for any GIS strategy to be successful.

6.1 INTRODUCTION

In all organisations, while the broad business directions and drivers are set 'at the top', the 'carrying-out' of those business directions is undertaken by 'line managers', and as obvious as it seems, the nature of most business unit line managers is to manage their business unit so that they meet their Key Performance Indicators (KPIs).

Therefore, a good business unit manager is most often primarily focused on his/her objectives and, in some cases, is not all that interested in assisting other business units to meet their objectives. This is exacerbated by competition between business units in some organisations, and 'helping the opposition' is not always at the forefront of the mind of a successful manager.

An outcome of this undercurrent of competition between business units in large organisations is a tendency for line managers to feel that 'this is my data – I've collected it with my hard fought budget, we maintain it, so why should I give it to you'. And then if they do provide it to other business units, or to the corporation as a whole, they often want

Achieving Business Success with GIS   Bruce Douglas
© 2008 John Wiley & Sons, Ltd   ISBN: 978-0-470-72724-9

something back in return, such as payment, increased budget, access to other capabilities, etc., which are not always forthcoming.

In this manner, the structure of an organisation can either help the GIS to be successful or severely impede its progress. As such, the structure of the organisation is absolutely crucial to whether the GIS will be successful or not.

6.2   IMPACT OF ORGANISATIONAL STRUCTURE ON GIS

If the organisational structure helps the GIS to access data which is corporate in nature but 'owned' by individual business units, then the GIS will have a higher chance of being successful. Conversely, if the organisational structure is such that accessing data (which is corporate in nature) is made difficult, then the GIS will almost always not be successful.

While organisational structure is not something that I wish to go into in any detail in this book (as there are a number of excellent books available on this topic), suffice it to say that a good organisational structure is absolutely critical to the success of GIS. In addition, most research on organisational structures generally highlights that the focus of any good (organisational) structure should be 'on outputs' rather than 'on inputs' – a concept which I would readily endorse.

Therefore, having said that a good organisational structure is absolutely critical to a good GIS, the organisational business units in the example cited in Chapter 4 are worth reviewing. If these units are appended with their major focus, it can be seen that there may be a potential conflict with a GIS (if implemented), i.e.:

- Operational Services – output focused.
- Corporate Governance – input focused.
- Finance – input focused.
- Engineering – input focused.
- Planning and Environment – input/output focused.

Note that only two of these five business units are focused on the customer; the other three are focused on the work going into the process, not out of it. Therefore, the employees are 'inward looking' rather than 'outward looking'. Their focus is on the ingredients in the recipe, not whether the cake being cooked looks good or tastes good when eaten. And as any cook will tell you, anyone can put all the ingredients together in a bowl, but very few can make a cake taste really good. Therefore, just as the cooking should focus on 'the cake' and not the ingredients, the business focus should be on 'the business outputs' and not the components which go to make up that business.

Hint

In any GIS Strategy and Implementation:

- the technology (and GIS) issues are 'the easy bits';
- the organisational issues are 'the hard bits'.

> This is because the organisational structure often does not facilitate the easy integration of data and information (spatial and aspatial) across an organisation, which is absolutely essential for a spatial environment to be successful.

In organisations such as these, a customer with a question will be directed to Engineering if it is an engineering question, to Planning and Environment if it is an environmental question, etc. Not only is this confusing to the customer and confusing to the organisation, more importantly (from a GIS perspective) it means that Engineering have to keep engineering data, Planning and Environment have to keep environmental data, etc., all in their own 'silos' and almost always not integrated and/or in conflict with one another.

A consequence of this type of structure is that it is difficult for a GIS to be effective because it will have to integrate data between different business units, and the biggest impediment to this happening will be the organisational structure, rather than the technical issues associated with the data or the technology.

Why is this important? For GIS to be effective in an organisation, it is generally used as a 'corporate tool' – therefore requiring access to data held across the organisation, often from different business units. Therefore, for a GIS to be successfully implemented, considerable effort must be given to:

- the creation and/or capturing of the required data;
- the maintenance of that data; and
- putting in place the organisational arrangements or structures which will support this concept and provide corporate access to the data and systems.

This is also not helped by many large organisations and government agencies continually reorganising their constitutant departments in the search for better organisational structures to meet their changing business goals. This then results in work departments, or parts of departments, being split to join with other departments or sections from other departments, all of which makes it difficult to maintain access to changing organisational silos.

Therefore, because organisational structures generally actively work against the successful implementation of a GIS, other arrangements have to be put in place in order to ensure that the GIS does get access to corporate data and can be used in a corporate manner.

6.3   ACHIEVING AN ORGANISATIONAL FOCUS FOR GIS

The normal method of achieving an organisational focus for GIS, without undertaking an organisational re-structure, is to ensure that there are the following:

1. *System 'Owner' or 'Custodian'*
   It is normal that an 'Owner' or 'Custodian' of the system be appointed from the business unit that has the most to gain (or lose) from the system being implemented successfully. Preferably this should be the business unit that has invested a considerable amount of their own budget in implementing the GIS, i.e. they have a strong business need to make this successful and will undoubtedly become the 'champion' of any system.

2. *GIS Coordinating/Steering Committee*

A committee of stakeholders from representative business groups who have an interest in the development and use of corporate data. This committee is usually formed to oversee and prioritise the implementation of the GIS and should consist of business unit managers, and *not* be delegated to technical officers. The system owner should be the chair of this committee.

Case example

A recent project involved reviewing the GIS for a major utility because of their very large and growing data backlog. It quickly became obvious that this backlog was increasing and the system was not meeting their business needs (they had changed their GIS 2 years previously from a 12-year-old heavily customised system).

One of the findings was that a major contributing factor as to why the new GIS was not meeting business needs was the lack of input by any 'real' business units in this process. The 'main players' for the last 2 years had been:

- the Data Capture/Conversion Team – they collected and converted the data but did not use the GIS or data for any specific business function, i.e. they were a 'provider of data' to their 'customer' business units;
- the GIS Support Team – they did not use the GIS or data, but were a 'provider of system support services' to their 'customer' business units; and
- the IT Project Team – they did not use the GIS or data, but were a 'provider of solutions' to their 'customer' business units.

Noticeably, the management model used did not include any business units who had the most to gain (or lose) from either the GIS system performing or the GIS data being correct, with or without a Service Level Agreement being in place.

Therefore, while each of the above groups had a positive attitude and 'all care' was taken, they were 'not accountable' to a business unit who wanted to use the GIS and its data, nor were they accountable to ensure that the GIS system performed as required or that the data was to the required standard – all of which equal a recipe for long-term problems.

---

3. *Data Custodians*

Data custodians should be nominated/appointed for particular datasets over which they have control. The custodians for such data are usually those individuals or groups who have the prime carriage of maintaining that data – i.e. their job function involves the continual update of that data.

4. *GIS Person/Team*

A GIS person/team will often be required to undertake the 'corporate' tasks of collecting and massaging data under the direction of the GIS Coordinating Committee, liaising with the various data custodians and providing support to the end-users.

In order to reinforce this structure, it is always a good principle that all datasets be allocated to data custodians and that all data custodians maintain their own data, i.e. data custodians should have 'write or edit' access to their data and read-only access to all other data. Note that this does not infer that the data custodian has to update the data personally – this can be delegated to the GIS work-group (usually under a Service Level Agreement) to do the actual work – the key issue here is that the data custodian has to *take responsibility* for the work being done, to give directions to the GIS work-group about what should and should not be done to the data. The GIS work-group must take direction from the data custodian.

If a custodian cannot be found for a dataset, then it should be seriously questioned as to whether that particular dataset should be captured or maintained. When this happens consideration should be given to whether that dataset should be 'struck off' the list and not be captured or maintained. This will almost always result in stakeholders focusing on whether this dataset is important to their business and, if so, whether they should nominate as the custodian.

To ensure that all data meets the corporate standards before being committed to the corporate GIS database, the GIS work-group should ensure that all data maintained by the data custodians is 'quality assured' to the standards of the organisation, 'catalogued' and contains a sufficient level of metadata prior to being committed to the database.

All corporate data should be able to be accessed by those staff whose privileges provide them with access to particular datasets, noting that some data (e.g. sensitive or confidential data) may not be made available to all staff.

In addition, in all GIS environments, there is often a need to categorise users as:

- Power users: for specific data entry and editing, map compilation, map presentation and analysis functions – power users are typically full-time (or close to full-time) GIS users and are usually highly skilled;
- Intermediate users: for ad hoc mapping and analysis tasks used probably several times a week; and
- View (web) users: for the remaining staff accessing a range of read-only data via a number of specialised applications (usually web based).

Each of these various categories of GIS use by staff will have different organisational impacts, particularly for training and support. Typically in a large organisation, the ratio of Power : Intermediate : View users would be of the order of 1 : 5 : 100.

6.4 BUSINESS PROCESS MAPPING AND RE-ENGINEERING

As stated repeatedly in this book, a good GIS Strategy must be based on meeting the business needs of the organisation. If this is undertaken correctly, the strategy should deliver substantial business benefit, most of which will be in the form of efficiency gains, achieved by having a reduced workload or by improved business processes.

However, some of the changes to business processes have the potential to be quite severe – in some cases so severe that the business process is no longer required, this being, as one would expect, much to the concern of the people employed to undertake that process. Therefore, prior to completing and implementing a GIS Strategy it is often wise to ensure that all issues relating to business processes and potential industrial relations problems are examined to determine if these issues need to be resolved before going further. A good method by which to accomplish this is to initiate a Business Process review to determine (or map) the processes which the GIS will probably impact, i.e.:

- the 'As Is' process, i.e. the process as it is at present; and
- the 'To Be' process, i.e. the process as it may appear after the GIS is implemented and 'bedded down'.

It is often useful to ensure that a Businss Analyst with substantial GIS experience is used to define the 'To Be' processes since this will require knowledge of the incoming GIS and how that capability will improve or change the business processes. This is the 'envision' stage in process review and one that is particularly important to 'get right'.

From analysis of the 'As Is' and 'To Be' processes, it can become apparent where the changes will occur and what data is required to support the 'To Be' process and, if required, to support any changes to migrate to this process. If properly implemented, the process should become more streamlined.

However, it is always prudent to ensure that such an analysis is cognisant of staff changes which might be brought on by streamlining business process, and indeed whether there will be any staff cuts as a consequence.

Case example

The implementation of a GIS and Document Management Strategy in a large Municipal Council resulted in radical change for a number of business processes which were based on the movement of paper around the organisation. The clerical people were, quite naturally, very concerned about their roles changing and in some cases their jobs being lost.

If technology is successfully implemented, there should be savings, and most often these savings are in the form of improved efficiency, that is, 'doing things better', often by doing them a different way. In this case, the paper process was replaced by an automated workflow which integrated the spatial and forms data in the development application process.

Considerable workshopping was undertaken with the staff whose roles were going to be changed in order to reassure them that their jobs were safe, but more importantly to engage them in the process and to have them define their changed roles. No staff were lost, substantial efficiencies were gained and a large amount of the paper process became automated with substantial savings to Council and faster turnaround for developers.

Typically, it is at this stage when one finds the most resistance from staff, as one would expect when staff start to realise that their job might radically change or disappear. Therefore, it is critical that this stage is handled very sensitively. It is also wise that senior management are actively involved at this stage and agree to reassign rather than retrench staff if positions are made redundant. This should be communicated to all staff regularly.

While Business Process Mapping is not something that I wish to go into in any detail in this book (as there are a number of excellent books available on this topic), suffice it to say that the successful implementation of GIS will almost certainly impact a number of business processes (this should be where most of the benefits are after all) and it is crucial to map these processes so that they can be re-engineered to take advantage of the GIS when installed.

6.5   TRAINING AND SUPPORT ISSUES

The provision of a competent, thorough and professional training and support programme is essential to the success of any GIS environment. Staff must have sufficient and context-sensitive training in order to undertake the functions required to use GIS and to do so in an efficient and diligent manner. It is also useful to augment this training by a programme of continual refreshing and mentoring to ensure that the skill level of staff remains at a 'peak condition'. This should also be complemented by a support programme to augment staff knowledge for all levels of users.

GIS TRAINING ISSUES

In the deployment of a major system with considerable technical content, it is usual and sensible for training to be focused and to be as effective as possible. As such, it is usual that training be available in the location required, at the skill level of the intending trainees and be context-sensitive (i.e. it must be relevant to the work processes of the organisation), noting that some users will require extended specialised training in specific subject areas.

It is typical that a training plan be developed which includes a component for the up-skilling of managers who need to understand IT in general and GIS in particular, with a particular focus on the need for data preservation in an 'information society'. A key component to developing this training plan is to ensure that the management of staff expectations is realistic so that the most appropriate use is made of geographic information.

GIS APPLICATION SUPPORT ISSUES

While the GIS is often located in a corporate area of an organisation, the management of the system should be with the 'system owner' and be supported by:

- IT and GIS staff for basic 'help desk' queries; and
- vendor staff (under a support and maintenance contract if required) for issues relating to the use of the GIS and related applications.

The support of any database applications (if required) may be best provided by either the organisation's IT Support or the provider of that application. The method of spatial system support reported in the GIS/Spatial Surveys is shown in the following chart. This indicates that vendor support is low and has decreased over recent years, while in-house support has increased at about 3% p.a. to 18%. Surprisingly, 1 in 10 contributors reported that they do not have any support for their GIS.

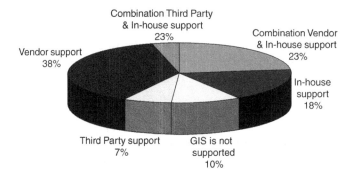

The level of satisfaction across all support provider groups is shown in the following chart, indicating that 90% of respondents would be continuing with their current support arrangements.

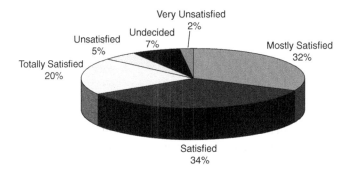

GIS DATA SUPPORT ISSUES

Again, because GIS is data-centric, a lot of support queries are often about the data rather than the applications using the data. Therefore, arrangements should be made with either the support staff to become up-skilled in data issues or that part of the support should be passed to the business domains/custodians whose roles are the management of that data.

6.6   SWOT ANALYSIS

One of the most common activities which is usually undertaken by an organisation contemplating a substantial change induced by technology is to do a SWOT analysis, i.e. to gather and analyse the Strengths, Weaknesses, Opportunities and Threats. A SWOT analysis can often show surprising results and is best undertaken in a workshop environment, where participants can leverage off comments made by others in the group to provide useful pointers to highlight impending issues.

While a SWOT generally focuses on 'the business', it can be undertaken on a number of levels, paticularly on the implementation of the GIS to highlight major areas of concern that may not have been collected from other interviews/workshops.

An output, or follow-on activity, from a SWOT usually is to undertake a risk analysis for any major IT or GIS implementation. Again, this is usually undertaken during a workshop, with each risk being identified and rated on its severity and criticality to the project at hand. During this process, it is also usual to identify how that risk can be mitigated and who will take carriage of that risk mitigation, etc.

6.7 SUMMARY – ORGANISATIONAL FOCUS

At this stage it should be apparent that in order to develop a GIS Stratgy it is first necessary to understand the organisation, to understand the business that the organisation conducts and to understand how the structure of the organisation can support, or in some cases impede, the business.

It should also be apparent that the organisational structure will be crucial to the successful implementation of a GIS. A good structure will support the notion of corporate data and a corporate GIS, while a 'less than satisfactory' structure will, in most cases, impede this. Unfortunately, it is not often practical (or achievable) to initiate an organisational restructure just so that the GIS can be accommodated.

Therefore understanding the organisational structural limitations and being able to deal with these limitations is essential. Having a good committee structure is vital, as well as an 'internal champion' who takes 'ownership' of the GIS process. If that internal champion is a senior manager, the probability that the GIS will be successful will be improved.

And, at the end of this, having a good understanding of the data and technical scope which may be required to implement a GIS system is very important. That is:

- good quality data is intrinsic to having a successful and effective GIS; but
- achieving good quality data will require overcoming organisational barriers.

An integral component of managing organisational issues is that it is essential to ensure that expectations of stakeholders are managed and that if there are any modifications to any business process the expectations are communicated carefully to all staff so that they are 'all on the same bus' and heading in the same direction. If this is ignored, then when organisational change results from a new GIS it is typical that the 'rearguard actions' will start to surface, to the detriment of the success of the implementation.

In summary, this chapter has discussed the need to ensure that there is an appropriate focus on the organisational issues involved in developing a GIS Strategy and in implementing the GIS. These issues are *very important* and should not be underestimated or ignored for any reason.

7 Developing the Application and Technology Focus

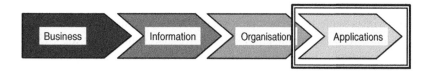

Lastly, and after developing the Business Focus, the Data Focus and the Organisational Focus, consideration can now be given to developing the Application and Technology Focus for the GIS Strategy. Unfortunately, this is often the point where many people start when considering GIS, consequently overlooking the business and organisational issues, much to their later peril.

7.1 GIS ISSUES

The following sections discuss some of the typical functional capabilities found in most contemporary GIS products. The requirements for these capabilities should be based on interviews undertaken with key stakeholders in the organisation. Such interviews usually highlight issues which have a business need and that may be able to be addressed by the use of geographic information technology.

GIS AS AN INFORMATION MANAGEMENT TOOL

A number of decisions undertaken by staff often refer to 'what', 'where', 'what are our responsibilities', 'how much' and 'how often'. The inclusion of the 'where' component into the decision-making process can be a powerful tool for providing a better understanding of the issues at hand and the implications of specific decision paths. A number of management staff in organisations may be able to access information presented geographically in order for them to be able to undertake their duties.

While it is typical that most staff will only need to view limited information themes and then only on an ad hoc basis, there will be a need for more sophisticated use by specifically trained persons, particularly by those who will be using the system for extensive periods. This often requires an easy-to-use, easy-to-understand and intuitive environment, accessible from desktop PCs, for users to view data, undertake rudimentary searches, undertake basic spatial/thematic mapping and constraints analysis and to print the resultant output. In some cases, this access could take the form of web-based applications.

Achieving Business Success with GIS   Bruce Douglas
© 2008 John Wiley & Sons, Ltd   ISBN: 978-0-470-72724-9

In this way, a geographic information management tool can be used to provide the means to locate information, as well as the means to provide better understanding of, and correlation between, different categories of information, such as engineering information, planning information, environmental information, etc. Such a management tool should be accessible to all staff across the organisation's network, be easy to use and provide comprehensive results.

CORPORATE DATABASE

While a GIS will provide the geographic/spatial basis for information management, a considerable amount of an organisation's users will also want to use the GIS to access and query non-geographic information relating to the business of the organisation – often assets, leases, permits, planning issues, etc. as well as data which may be stored in the corporate financial system.

In addition, it is often found that a considerable amount of information of this type is currently stored either in Microsoft® Access databases or on paper. Because Access databases can be developed with minimal technical skills and the software is often included on all desktops, they are often used extensively. It is common to find lots of people developing lots of database applications which store lots of data in lots of different (and unique) manners – often duplicating data stored in other similar databases.

Case example

A recent GIS Strategy for an organisation of 70 staff using 55 PCs revealed that there were 533 Access databases resident on the network (i.e. an average of almost 10 databases per user). This very large number of databases suggested that:

- staff were spending an inordinate amount of time generating Access databases to undertake tasks which should be able to be undertaken more efficiently if done in a coordinated and corporate manner; or
- a large number of databases were not used and could be archived/deleted; or
- there were multiple versions of the same data in multiple locations without anyone having an understanding of which dataset (if any) was correct or the most recent.

Huge amounts of time were thus being wasted and large amounts of data were being duplicated across different databases with a consequent deterioration of data integrity to the point where users could get almost any answer they wanted by using different databases.

While Access databases are often considered to be the 'bane of the life' of an IT manager, they are a 'fact of life' for most organisations and do serve a purpose (if properly developed and managed). Experience has shown that it is often better for an organisation to adopt an 'embrace and manage' attitude for Access database developments rather than to ban them and have users consequently going 'underground' to use Excel as a pseudo-Access environment.

In order for the organisation's users to use the information in Access through the GIS (and in order to derive the benefits from having a GIS), it is often logical that these Access databases be rationalised and that the GIS be augmented by a corporate database which can store the data that is non-geographic (textual) in nature.

This type of corporate database will require the development of a corporate data model (and data dictionary) to ensure that the database meets the wider corporate information needs of the organisation.

BASIC PRESENTATION AND OUTPUT REQUIREMENTS

Most GIS products categorise spatial data on a 'layer' or 'theme' basis, display data in combinations of layers, add attributes to the spatial data, annotate data, colour lines/polygons and add line-styles and symbology to create a representation of the 'real world' being mapped.

The GIS should be able to produce high-quality map and screen output, including presentation graphics, traditional mapping output, map products developed on an ad hoc basis (e.g. clip of current map screen onto a generic map-sheet border) and map products at a specific scale and map sheet area.

It is always useful for the GIS presentation tool to include the intelligent treatment of text sizing, text positioning and placement of other cartographic features (e.g. symbology) based upon the output scale and map theme selected, particularly if a semi-automatic thematic map-based presentation tool capability is used.

There is also a need for the GIS to be able to display information when specific 'zoom scales' thresholds are reached so that more information is displayed as one zooms into a map. This is usually applied on a layer basis, i.e. layers are turned on and off as zoom scales are reached. This also avoids the cluttering effect apparent when huge amounts of the data are in the database, but the display scale is such that all of this data is only shown 'as a blob' unless some form of data filtering is applied.

PROJECTION MANAGEMENT

Because a GIS contains data which is a physical representation of the surface of the earth, the result is a 'round' rather than a 'flat' system. Data is referenced to a real-world location such that a specific piece of 'where' data inherently knows the location of all other 'where' data, and vice versa. That is, it is not necessary to establish explicit relationships between 'where' data because the 'system' intrinsically 'knows' the relationship.

The map base inherent within a geospatial system is a mathematically defined representation of the surface of the earth. Defining the earth's surface is a specialised branch of surveying and, while there are numerous models that define the earth's surface, each with their own set of mathematical properties, the important aspect is that once a particular model has been selected all 'where' data that is input to the system thereafter must be expressed in terms of that model. These models are based on geodetic datums and use projections to transform 'round' data onto 'flat' paper. Some of these projections are more suited to smaller geographic areas while others are more suited to mapping a whole country, therefore it is important to ensure that data used from another system is on the same datum and uses the same projection as the data inherent in the host system.

It should be stressed that the map projection/datum reference systems are fundamentally important to the quality of spatial data and should be included in all metadata. Having the wrong coordinate reference system may not be immediately apparent to the untrained user but could result in the spatial data being mis-located, in some cases by just enough to not be obvious, or else by a substantial amount. In some cases, having the wrong coordinate reference system information can be worse than having no coordinate reference system shown at all.

Therefore, the GIS should be able to use a number of map projections/datums and be able to convert data 'on-the-fly' from one projection to another. All of the datums and projections applicable to that country and/or region should be included in the GIS product and should be supported by the vendor. This should also include the ability to store spatial data with geographic coordinates (i.e. as Latitude and Longitude) if required.

POLYGON AND LINEAR ANALYSIS

There is often a strong need within organisations for the GIS to be used for creating and analysing polygon and linear objects and for consideration of buffers along those polygons or specific corridors (e.g. within 10 m of a high-pressure pipeline) and features of environmental interest such as groundwater (e.g. for understanding the impact of a possible contaminate). This should also include the ability to be able to undertake modelling of polygons, i.e. to add/subtract polygons and to create new polygons for subsequent analysis.

NETWORK MODELLING/ROUTING

A key requirement for some organisations, particularly utilities, is the ability to trace a linear network, such as a telephone/water/power network or a road network, to be able to check the integrity of the connectivity of the network and then to be able to 'map catchments' of these networks for subsequent analysis. Catchments in this context may be a power zone, a water supply network or a suburb. Based on this network tracing and analysis of catchments, the organisation may then wish to undertake further analysis using specialised packages to determine, for example, power supply outage, hydrological modelling, etc.

Because these requirements are very specialised and relate to the management of linear networks, as distinct from the management of polygon networks, the software requirements are quite different and very specialised. A number of GIS applications specialise in network-based systems.

As well as network analysis, some organisations also wish to be able to integrate the functionality of network-based systems with that of polygon-based systems, particularly where there is a need to manage linear assets (e.g. pipelines) in the context of polygons (e.g. customer's property, parks, etc.).

IMAGE UNDERLAY AND ANALYSIS

Most GIS systems make extensive use of imagery to provide a very useful qualitative and contextual background which can be used to identify features or from which to digitise items or boundaries visible in the photography. However, to do this the imagery must be rectified so that all of the distortions inherent in aerial photography are removed and the image is at the correct geographic location and orientation. This photogrammetric conversion of aerial photography to (digital) orthophotography will result in a correct 'map-accurate' base for the GIS (subject to image resolution and positional accuracy).

Because imagery contains a wealth of descriptive information, it can be used as a valuable data source to undertake a range of business activates. It is also particularly useful when 'draped' over a 3D terrain surface and used as a visualisation/analysis tool.

Commonly used imagery is typically orthorectified aerial photography or remotely sensed imagery (either aerial or satellite based). Remotely sensed imagery has the potential to analyse spectral bands (if the software is available) to focus the analysis on specific issues, e.g. crop disease, and when used as such it can be a very powerful tool for broad-scale analysis and planning.

But imagery should not be limited to just landscapes of aerial photography or satellite imagery. GISs often have a need to underlay scanned images of hard-copy drawings/maps (e.g. Civil Drawings) which may cover part of the geographic area in question. In addition, it is often useful to attach images (such as photographs) to features/assets in a GIS so that all the information about that particular data item can be accessed. For example, a photograph of a pump attached to the pump asset in the GIS can convey substantial descriptive information which may not be able to be stored in attribute tables.

MAP/TEXT/DRAWING HYPERLINKING

A lot of information used by an organisation and provided to internal and external stake-holders often consists of textual information, maps and charts/tables. However, in an 'on-line' information society there is a need for this information to be available in an interchangeable and easy-to-use manner. For example, while an enquiry for leasing may commence with the identification of a location on a map, it may quickly move to the need to determine information about lease conditions, constraints, development conditions, etc. – most of which may be available in another form (e.g. as text or as a drawing).

Some of this information may be available as CAD drawings which can relate to a map location but they may not be in a form which can be easily integrated within the map. Therefore, it is very useful for the GIS to be able to embed hyperlinks which enable the information to be retrieved from other (often non-spatial) databases for presentation to the user.

3D MODELLING AND ANALYSIS

A number of organisations often have a need to be able to store, view and analyse data in three dimensions. While most GIS applications are based upon data being stored and managed as 2D, the third dimension can be extremely useful to undertake a number of modelling and analysis tasks. Usually this is undertaken by specialised software as an additional module to the core GIS application.

However, careful consideration should be given as to whether there is a legitimate need for 3D functionality or not. 3D software is expensive and is often purchased but not used, except in the initial implementation when the expression 'gee, look at that' can be heard throughout the office.

Often, the need for 3D can be taken off-line to a specialised application (e.g. for Surveying/Civil Engineering and for Mining Engineering applications) where environments of this type often provide substantially more flexibility with less 'geo-processing' overhead in order to undertake complex analysis in a timely manner.

It is also worth stating, as obvious as it seems, that 3D applications require 3D data. The data storage overhead, as well as the cost of acquiring data with a third dimension (usually, but not always, height), can be substantial and should be carefully weighed against the business need, i.e. will this data be used for any practical business purposes?

It is also worth noting that the 3D/4D (3D + time) software 'genre' is rapidly changing – systems are currently being built based on games technology to provide an animated 'games' look or feel which often needs ready access to huge volumes of 3D data in real-time. When that is combined with advanced image processing techniques to allow image 'draping' across 3D landscapes while travelling through the landscape, the results can be amazing. However, these applications are high-end, very specialised and often require very highly optimised (and expensive) hardware.

TRANSACTION MANAGEMENT

A number of GIS systems have a 'transaction management' capability which relies on a 'job' being opened, actions being undertaken, the job edits being 'posted' to the database and then the job being closed.

In an operational environment with up-to-date data and with a number of users accessing data assets in real-time, this is a valid, competent and logical process which should result in data of high integrity where chronologically-based business rules are observed. That is, a 'job' is opened, edited, posted and closed, and then at some future time another job is opened at the same (or nearby) asset, edited, posted and closed.

An extension of this concept are 'long transactions', given that some jobs may be open for an hour, a week or occasionally a month. Typically a 'transaction log' is used to manage this process.

Occasionally, a conflict may occur when two or more users try to post edits on the same feature. When these job edits are 'posted', the first edit posted will be accepted and subsequent edits will raise conflicts which the editor then has the option of accepting or discarding. When there are more than two edits undertaken on the same feature (i.e. because there are more than two jobs involved), a conflict will occur which may have to be resolved by the GIS operators.

While some GIS systems lock the data records (to the first user) so that others cannot edit the data while the first user 'has the data', other GIS systems do not do this – preferring instead to resolve the conflict at the end of the process, rather than locking the second or third user out of the process altogether. The notion of whether to lock users out at the start or end of the process is often a philosophical debate, but both concepts have their advantages and disadvantages.

However, a consequence of this philosophy is that maintaining proper chronology is extremely important. All data must be processed with chronology in mind, all processes (transactions) must have a 'start' and an 'end' and, while 'long' transactions are possible, it is expected that there are not too many 'long' transactions or that they do not last 'too long'. That is, the user and system manager must work within the context of this product philosophy.

DYNAMIC SEGMENTATION OR LINEAR REFERENCING

Some business applications require that the GIS be able to store and segment linear data in real-time so that various combinations of attributes or features can be extracted and analysed. The typical business application of dynamic segmentation is usually in road/highway management.

An example of dynamic segmentation for a road management application would be as follows. A requirement may be to analyse a segment of road which has been categorised as having:

- surface cracking/rutting attributed between 2600 and 3500 m;
- narrow shoulders attributed between 1800 and 2900 m;
- the road may be a single-lane carriageway (stored spatially) between 2200 and 9300 m; and
- re-surfacing attributed as having been undertaken between 500 and 2300 m.

An analysis to show all road segments that are single-lane carriageway, that have not been re-surfaced, that have narrow shoulders but with no cracking/rutting should segment the data (some spatial and some attribute) in real-time (i.e. dynamically) and return a result showing that only road segments between 2300 and 2600 m meet these criteria.

Dynamic segmentation often requires that the data be heavily structured in the form required for the specific GIS system and, as such, is a data overhead which should be carefully considered before being implemented.

WEB MAPPING CAPABILITY

Most contemporary GIS products include an Internet map server capability designed to publish spatial data to the Internet or intranet using generic web browser technology such as Internet Explorer (IE).

These types of products are typically installed on a separate web server, either internal or external to the organisation's firewall depending on the architecture implemented, and typically have the ability to deliver dynamic maps and GIS data and services in real-time to a wide range of users. This functionality can often be deployed for almost no cost (after the application has been developed) since web mapping is typically based on IE browsers. These web mapping applications are usually very easy to use and usually meet the needs of corporate intranets as well as the demands for Internet access.

Typically an organisation would implement a web mapping capability containing a customised application based around their business needs. Some examples of customised web applications might include:

- Find/View application – to generate a map at a specific location showing required data themes at a required scale.
- Routing application – to generate driving directions and route maps between different locations.
- Planning application – to retrieve all relevant planning and related information so that appropriate decisions can be made.
- Real Estate application – to locate a home that matches specific search criteria and provide a neighbourhood view (perhaps including aerial imagery) with a report.

As the use of GIS matures in many organisations, it has been found that the demand for web mapping capability increases, often in direct proportion to the falling demand for other GIS client applications.

PDA/GPS CAPABILITY

Most organisations have workforces which are reasonably mobile, and just as the humble telephone has become more mobile, so too are GIS systems, particularly where there is a need for data collection/update in the field. A consequence of this is that it is now common for GIS to include a PDA capability for use by staff when mobile.

The GIS/Spatial Best Practice Surveys have found that there are six specific uses of mobile devices (see the following chart). The most prolific uses of this technology are data collection (for input/update of office-based systems) and query access to data while in the field (54% collectively), while more operational uses account for the remaining 46%.

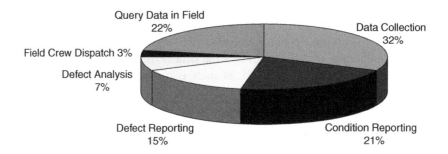

It is typical that PDAs and other mobile devices include a GPS capability (e.g. satellite navigation) to provide instant locational information as well as a 'cut-down' version of the GIS and available data for use in the field for a range of general purposes such as identification of location, i.e. 'where am I and whose property am I on' and 'what underground services are located near my location', etc. In addition to this 'generic' mobile GIS capability, it is also common for a number of small applications to be developed to focus the mobile workforce specifically on the set of field data collection/update tasks undertaken by specific staff.

TEMPORAL ANALYSIS

Most data changes over time and, by analysing these changes, trends and patterns can be determined which have the potential to extend conventional data analysis techniques and methods to deal with such temporal dependencies. For this to occur the GIS needs to store historical geometric and attribute configurations that can be overlaid together to facilitate comparisons, trend analysis and error interpretation.

As such, the organisation may need the ability to visualise and analyse complex time series within the GIS, since it is very difficult to understand complex dynamic systems like pollutant spreads, flood events, fire scaring, etc. without a temporal component.

INTERFACES AND INTEGRATION WITH OTHER SYSTEMS

In the maturing technology environment of many large organisations, it is becoming more common for GIS to interface and/or integrate with other corporate systems, particularly to facilitate access to data which may be common across such systems.

The level of integration will be unique for each GIS environment and be focused on the systems which are in use by the host organisation. Typically these systems would be Asset Management systems, Financial Management systems, CAD systems and other specific-purpose Engineering systems.

Integration is discussed further in Section 7.3 below, particularly with a view to data sharing and forming data alliances.

GIS REQUIREMENTS SUMMARY

In summary, the GIS requirements for organisations are quite variable and will change depending on the business of the organisation and the maturity of that organisation to make use of IT/GIS.

There are a number of GIS products available on the market which would suit the needs of any particular organisation. As indicated earlier, there is no 'best' GIS system – in much the same way as there is no 'best' motor vehicle. It all depends on the intended use, the available budget, the purpose to which the GIS is proposed to be used and the ability to make the best use of any available data.

While the above GIS issues are generic and common across most GIS systems, some systems focus on particular capabilities such as linear network management, whereas others may have capabilities directed at other functionalities such as business intelligence, web mapping, etc. Therefore, while all systems are different, the procurement of a GIS should be focused on selecting a system which meets the functional requirements of the business so that, once implemented, the system meets, and indeed exceeds, those business requirements.

At the end of the day, GIS is an information management tool, and if it is properly implemented it should be 'just another information management tool' on the user's desktop, and in this sense GIS should become as endemic as the use of word processing or email capabilities.

7.2   IT ISSUES

The implementation of a modern GIS in an organisation can often be impacted by the IT strategy of that organisation. A decade ago most GIS systems were stand-alone environments (often Unix) isolated from the corporate IT environment and therefore not impacted by any strategy that the IT environment was pursuing.

However this is typically not the case with contemporary GIS. Almost all GISs are now implemented as part of the corporate IT environment, running from corporate servers and being accessed by Microsoft® Windows-based desktops which also run a range of other corporate applications, such as Financial systems and Office systems.

As such the IT structure, corporate database, desktop configuration, operating system, network performance and Internet access are all issues which need to be considered by a corporate GIS. These issues are discussed below.

IT CONFIGURATION

Typically most business will be running a contemporary Microsoft® Windows operating system. This has been reported as being 76% of usage and growing at between 5% and 9% p.a. in recent surveys. Noticeably, the reported usage of Unix has reduced almost linearly from 24% in 2000 to being only 3% of usage at present (see Chapter 1). Conversely, Linux has risen from almost zero to just over 2% in the last couple of years.

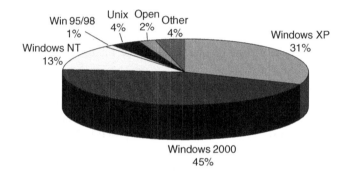

Most IT environments will consist of a number of servers, networks and desktop PCs deployed to deliver corporate applications to the desktop. It is now common for all office employees to have access to a computer and to be able to access corporate applications. Most large organisations 'roll-out' a Common or Standard Operating Environment (COE/SOE) across all desktops containing 'standard' applications on each desktop. If GIS is to be used as a corporate application, then it is typically implemented as part of the desktop SOE for that organisation.

DESKTOP CONFIGURATION

When developing a GIS Strategy, the desktop configurations should be reviewed in order to determine the ability of this platform to support GIS access, in particular the software suite on the organisation's deployed SOE. In particular, compatibility of the GIS genre (which may be selected) with the organisation's operating systems should be considered, as well as the configurations of the desktops used by the power users' with consideration of the need for upgraded memory, disc storage, etc.

NETWORK PERFORMANCE

The performance of most GIS environments will suffer if there is a need to deploy the technology across a LAN / WAN (Local Area Network / Wide Area Network) of limited bandwidth, particularly if there is a need to have a centralised architecture (e.g. Citrix) which requires the GIS application to continually access data (and/or the application) held on a distant server, and particularly if the data is stored as graphics files.

In some cases the network strategy, and the consequent network performance, may suggest that the most suitable GIS architecture should change to accommodate poor network performance. Conversely, an upgrade of the network to higher-speed lines (e.g. fibre optic or microwave) may be required to support a different GIS architecture. This is because most GIS applications are notoriously high users of any available bandwidth – moving graphics files between servers and the users often takes some considerable time and often adversely impacts other data on the network, particularly emails with large attachments (see Section 5.3, which discusses the different data architectures and the impact that different system configurations could have on the data architectures of the GIS).

CORPORATE DATABASE

Almost all large organisations have adopted a consolidated approach to storing 'corporate information' in an organisationally centralised database, often based on the adoption of Oracle, DB2 or SQL Server as a RDBMS used as the repository for data storage and accessed by corporate applications such as Finance, Human Resources, Rates, etc.

The RDBMS used by GIS users for the storage of attribute (not spatial) data is shown in the following chart, indicating that apart from data being stored in a vendor proprietary format, Oracle and SQL Server are the two most commonly used databases for the storage of corporate information accessed by GIS.

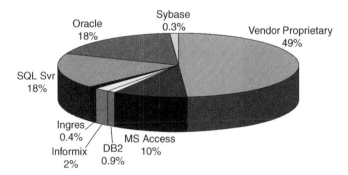

While each database product has specific advantages and disadvantages (in much the same manner as each GIS product has specific advantages and disadvantages), the use of a corporate database provides the ability for some degree of integration of data between applications, even if it is only through being stored in the same database and able to be accessed in a known format. For these reasons, the corporate database used by an organisation will impact onto the development of a GIS Strategy. Note that most GIS products have a preference for working with specific RDBMS products.

7.3   SYSTEM/DATA INTEGRATION ISSUES

In the maturing technology environment of many large organisations, it is becoming common for the GIS to interface and/or integrate to other corporate systems, in order to ensure that the data used by the major systems within an organisation can be re-used by other applications and to ensure that there is a tighter level of use of the technology across the organisation. The GIS Survey found that the systems reported to be integrated with GIS were as shown in the following chart.

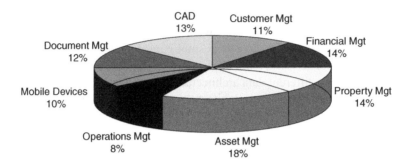

Integration levels can often be an indicator of the level of maturity for that industry regarding the extent to which spatial system functionality is being harnessed to assist with other areas of the business and hence to assist in meeting business goals and objectives, or to provide competitive advantages.

In Australia and New Zealand, the GIS Survey canvassed integration to the following eight major system types, often considered to be the major systems within organisations that would have a need for a GIS:

1. Customer Management
2. Financial Management
3. Property Management
4. Asset Management
5. Operational Management
6. Mobile Computing Devices
7. Document Management
8. CAD (Computer-Aided Drafting)

The distribution of the level of integration of spatial systems with other systems across all reported systems is shown in the following chart.

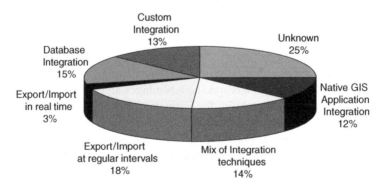

Sections 5.5 and 5.6 discussed 'GIS Data Standards and Related Issues' and 'GIS Data Interoperability', respectively, indicating that considerable work has been undertaken for a number of years, and is still being undertaken, to address the vexed issue of exchanging data between systems, whether it be by integrating systems or integrating the data flows from these systems.

As such, the integration of major corporate systems in many organisations is often focused on exchanging data, generally by creating an environment where there is access to a corporate 'data store' by many applications, or by tightly integrating applications which need access to common data.

The above chart shows that 15% of system integration involves integration at the database level, with most of this integration being the storage of spatial data in a relational database such as Oracle Spatial, where the spatial data is divorced from the spatial applications which may be used to edit and create that data. In this manner, different applications can access a single database of spatial information with the potential to provide the ability to exchange data between these applications without that data being 'hard wired' into the application. This type of integration then negates the need to move data from one system to another, in that it is stored in an independent database accessible from two or more systems. Section 3.3 discussed this in the context of 'Disruptive Technology'.

In addition to these technical issues, any discussion on system/data integration should include the need to form alliances with other 'like' organisations (e.g. government agencies) who may have data required by the other organisation. It is through such alliances that discussion will yield more than just the exchange of data – the development of common naming conventions and levels of data attribution will almost certainly provide a higher degree of usefulness of the data for all participants.

But beware, this process may also lead to the data being described at a 'lowest common denominator' level, and not being fit for the purpose required by either organisation.

7.4 DEVELOPING THE FUNCTIONAL REQUIREMENTS SPECIFICATION

Analysis of the GIS and IT requirements (as discussed above) typically forms the basis of a Functional Requirements Specification (FRS) which outlines (as one would expect) the functions that the GIS will have when it is implemented. Included in the FRS is also further detail about the data architecture, data types and the impact that the data has on the overall system functionality.

Therefore, the FRS is typically a specification containing data issues as well as application/infrastructure issues. The FRS should be a reasonably detailed document, based on the relevant functional categories highlighted above and documented to a level of detail that completely specifies all of the functionality required in the GIS, as well as ancillary requirements such as interfaces, database requirements, application development environments, etc.

The application/infrastructure component of the FRS is also typically used to form the basis of a tender specification (if required) from which to request tender responses, including detailed costing information, from vendors.

Similarly, the data analysis component of the FRS can be used as the basis of another detailed specification which may be required to tender for the capture/conversion of particular datasets as required.

In summary therefore, the FRS is developed after a reasonably detailed analysis of the applications (GIS)/infrastructure functionality and the data/information requirements that are required to deliver the outcomes (particularly the benefits) shown in the GIS Strategy.

That is, when the GIS Strategy is approved and implemented, the GIS system should meet all of the functionality contained within the FRS.

Note that the FRS should not be written to focus on the capability of a specific vendor. The FRS should be focused on the functionality required to meet the business needs, with the vendor capabilities subsequently evaluated against those business needs.

Should the subsequent tender evaluation find that the selected GIS cannot meet all of the requirements shown in the FRS, then the decision can be taken to either forego that functionality (and the resultant business outcomes that it may have provided) or to have additional software developed to meet that specific requirement. That is, the functionality required for the business should drive the capability delivered by the vendors, not the other way around.

The following extract from an FRS for several areas of functionality provides an indication of the level of detail typically provided:

x.8.   Three Dimensional Functionality

x.8.1   *The GIS should provide the ability to create a Digital Elevation Model by extending plane coordinate data into three dimensional data, by importing three dimensional data from external sources, merging height data with plane coordinate data or by forming three dimensional data from two dimensional contour data where specific contours are attributed with the height of that contour.*

x.8.2   *The Digital Elevation Model should be able to be created by extending plane coordinate data into 3D data using an attribute value on the plane coordinates which may not be height related but may be population density, cost, etc.*

x.8.3   *The 3D capability should be able to view multiple Digital Elevation Models which can be selected and loaded by users.*

x.8.4   *The 3D capability should be able to return the height from selected Digital Elevation Models from a screen digitised cursor location.*

x.8.5   *The 3D capability should be able to provide an Isometric view of a Digital Elevation Model.*

x.8.6   *The 3D capability should be able to provide a Perspective view of a Digital Elevation Model with the ability to set the eye location and the vanishing point, etc.*

x.9.   Raster Image Integration and Viewing

x.9.1   *The integration of raster images in the GIS will be used to enable orthophotography (rectified aerial photography) to be able to be used in the same environment (backdrop) as the vector GIS functions as well as to enable correlation of other image types.*

x.9.2   *The GIS system should have the ability to store, view and manipulate raster and vector data in the same environment.*

x.9.3   *The GIS should have the ability to display multiple, overlapping aerial photography containing different resolutions as a backdrop to the map display and support the following formats – TIF; GeoTIFF; JPEG; JPEG2000; BMP; GIF; ER Mapper ERS; ER Mapper ECW; MrSID.*

x.9.4   *The raster data should be able to be stored either as continuous grey-tone or in full colour.*

x.9.5   *The GIS system should make use of recent advances in 'wavelet' technology in the storage, compression and viewing of raster images as viewing scale varies.*

x.9.6 *The GIS raster image capability should be able to access large volumes of image data without significant degradation of system performance.*

x.9.7 *The image should be able to be calibrated or rectified onto existing database features or to specific coordinates for reconciliation with the vector data. This rectification should use an appropriate calibration (calibrations supported by the solution to be stated), and be able to make use of a large number of control points (at least 10) with suitable calculation and display of residuals, etc. This calibration should be able to be stored for subsequent re-use at later work sessions.*

x.9.8 *The GIS should be able to view TIFF, JPEG, GIF and PNG files in a separate image viewer.*

8 Developing a GIS Strategy

Chapter 3 highlighted that a GIS Strategy must be focused on the following four elements:

1. Business issues
2. Data/Information issues
3. Organisational issues
4. Application and Technology issues

In some cases the Application and Technology issues are split into two to provide more definition between the application and database software and the hardware and network infrastructure which supports it – however, the end result remains the same.

Each of these four elements was discussed separately in previous chapters and can now be brought together to form the basis for abstracting the GIS into the business environment as the last step in developing the GIS Strategy specific to that organisation.

8.1   FUNCTIONAL REQUIREMENTS SPECIFICATION (FRS)

As discussed in the previous chapter, the analysis of the GIS and IT requirements typically form the basis of a Functional Requirements Specification (FRS) which, when coupled with the data/information requirements, can be used to specify all of the technical requirements necessary for the GIS environment.

As such, the FRS becomes an integral component of the GIS Strategy, with the functionality specified in the FRS delivering the outcomes which are shown as benefits in the GIS Strategy. That is, there must be a one-to-one correlation between the intent and content of the FRS and that of the GIS Strategy. Put another way, there is little point in listing benefits in the GIS Strategy if they cannot be delivered by the functionality/data capabilities shown in the FRS.

Again, it should be stressed that the FRS should *not* be written to focus on the capability of a specific vendor. The functionality and data requirements, as described in the FRS, should drive the capability delivered by the vendors, not the other way around, albeit that some vendor capability (e.g. mobile applications) may be used to 'seed' discussion on business requirements.

Typically the FRS then becomes the technical requirements of a subsequent tender document should the GIS Strategy proceed to procurement. This is discussed in more detail in later chapters.

However, for the development of a GIS Strategy, the FRS has a specific purpose – to form the basis for ensuring that it has a high degree of 'business fit' with the needs of the organisation, as shown by a correlation with the Critical Success Factors (CSFs).

Achieving Business Success with GIS   Bruce Douglas
© 2008 John Wiley & Sons, Ltd   ISBN: 978-0-470-72724-9

8.2   CORRELATING AGAINST THE CSFs

Chapter 4 discussed the issues which need to be undertaken in the development of the business focus – the first element of the GIS Strategy. This process is typically undertaken by a series of interviews/discussions with management staff in the organisation in conjunction with a review of the documents which form the strategy of the organisation.

The outcome of this process is a broad set of CSFs which succinctly describe the business goals and objectives of the organisation, firstly at a high level and then subsequently at the more detailed levels of the component business units of the organisation.

It is at this stage that the CSFs can be correlated against the broad topics contained in the FRS to highlight specific technical areas where there will be a high business gain and those areas where the gain may be less. As such, this correlation can provide a good indicator for where the focus should be placed. Additionally, this correlation would also highlight functionality (if included) that may not have any business focus and can therefore be omitted.

The chart below is one of (typically) a series which shows the correlation, expressed as High, Medium or Low, of the requirements (as documented in the FRS) against the business drivers, shown as the CSFs. As can be seen, the requirement to 'Develop a Corporate Database' has a High impact on four CSFs whereas 'Implement Metadata Manager' has a High impact on only one CSF but is necessary to support several other functional requirements. In addition, it is apparent that CSF 1, and CSF 2 to a lesser extent, will not be impacted by the GIS, whereas CSFs 3, 4, 5 and 7 will all be substantially enhanced should the GIS be implemented.

Critical Success Factors

1. Increase business operations and maximise financial and economic return
2. Provide level of infrastructure to support business operations and outcomes
3. Reliable, timely, accurate & integrated data to support infrastructure & services
4. Minimise risk by data being correct, up-to-date and available
5. Trained & efficient staff using appropriate tools
6. Sustainable growth through better environmental outcomes
7. Timely response to community needs

Requirements — Correlation of CSFs and User Requirements

Requirements	1	2	3	4	5	6	7
Organisational							
Create GIS Work Group	Low	Low	High	High	High	Medium	High
Develop training programme	Low	Low	High	High	High	Medium	Medium
GIS & IT Infrastructure							
Implement GIS software	Low	Low	High	High	High	High	High
Develop Corporate Database	Medium	Medium	High	High	High	Medium	High
Develop interface to Financial system	Medium	Medium	High	High	Medium	Medium	High
Acquire / develop Lease Mgt system	Medium	High	High	High	High	Medium	High
Acquire PDA capability	Low	Medium	High	High	Medium	High	High
Implement GIS intranet viewing	Medium	Medium	High	High	Medium	Medium	High
Implement Metadata Manager	Low	Medium	Medium	Medium	Medium	Medium	High
Purchase additional hardware	Low	Medium	High	Medium	Medium	Low	Medium
Data							
Capture aerial photography	Medium	High	High	High	Medium	Medium	High
Capture underground infrastructure data	Medium	High	High	High	Medium	Medium	High
Collect / convert external data	Low	Medium	High	High	Medium	Medium	Medium

This type of correlation highlights to senior managers in a very easy to understand form that the GIS is focused on the objectives of the business and, if implemented, will substantially help the organisation to meet the business objectives.

8.3 DEVELOPING THE GIS STRATEGY

Once the FRS and CSFs have been defined and correlated, the GIS Strategy can then be put together as a series of recommendations with attendant timelines and 'roadmap' to implement the outcomes. That is, the GIS is abstracted into the business environment to deliver the benefits stated in the Strategy.

As indicated previously, the GIS Strategy for an organisation is uniquely developed for each organisation in order to meet:

- specific business drivers;
- specific data requirements (both current and required data);
- specific organisational requirements; and
- specific application/infrastructure requirements of that organisation.

As such, just as every organisation is different, so too is each GIS Strategy.

8.4 SUMMARY

In summary therefore, it is critical that the business and organisational issues are addressed before any technology considerations are undertaken.

If the business and organisational issues can be 'sorted out' to facilitate the creation of a corporate approach to data and to break down the 'data silos' inherent in most organisations, then the GIS will have every chance of success. When these business and organisational issues have been addressed, the data issues most often 'fall into place' and result in developing procedures so that data custodial and access protocols will (ultimately) support the business goals.

GIS software and the supporting IT environment is usually the least concerning issue but the one which often receives the most attention, since it is also the one that is often the most easily understood and solved. This is also the direction that the product vendors are 'coming from' and so, in their minds, it is the most important issue (for them) – hence it is the area where considerable discussion occurs.

A GIS Strategy is typically 'rounded-off' by the inclusion of:

- a broad implementation timeline;
- the outcomes of the cost/benefit analysis (see next chapter);
- an architectural diagram highlighting how the GIS will 'mesh' with the existing IT environment when implemented; and
- recommendations and action plans for the next steps.

9 Cost/Benefit Analysis/Return on Investment

Major initiatives in most organisations, particularly those involving the expenditure of large sums of money, generally need to be justified and approved by senior management. This is particularly important for IT projects, with industry experience indicating that over 70% of all IT projects fail, some with disastrous consequences.

This need for accountability and high-level approval of budget commitments also applies to GIS, in some cases warranting a more rigorous financial assessment, often due to a lack of clarity and understanding of the benefits that GIS can bring to an organisation. As such, almost all organisations will require that a financial analysis be undertaken and that this be in the form of a Cost/Benefit Analysis and Return on Investment calculation. In some cases, this is also extended to the need to develop a Business Case for presentation to Executive Management or the Board. These topics are discussed in this chapter.

The Cost/Benefit Analysis is usually developed to show all of the costs and potential benefits for the proposed GIS and to highlight whether there is a Return on Investment (ROI) over (typically) a 5-year period. The Net Present Value (NPV) should also be calculated using several interest rates (to show the sensitivity of interest rate fluctuation to the NPV). Five years is generally recommended as the maximum period of time to undertake such an analysis, as technology 'turns-over' at least every 5 years and a longer period is often difficult to sustain from a technology perspective. Equally, a period shorter than 5 years often does not allow sufficient time to gain a return from the system.

Unfortunately some GIS managers consider the process of developing a Cost/Benefit Analysis as a process to be endured (often with some reluctance) rather than considering it as a useful exercise to ensure that all the facts are at hand and presented in a full and frank manner to management, along with all issues and potential risks documented.

Therefore when developing a Cost/Benefit Analysis, it is essential that it be done in such a manner that it is understandable by senior management and has a focus on the business of the organisation. This can often be a painstaking and complicated process.

This is particularly important for GIS, because most GIS failures are because:

- the GIS project was not 'thought through' properly and/or not planned;
- the GIS was 'over-sold and under-delivered' by the internal project team;
- the internal project team believed the vendor when he/she said 'trust me, etc. etc. . . . .';
- while the GIS data was required from across the organisation, the data was in 'locked-down silos' and not available to the GIS, thereby limiting most of the benefits envisaged;
- the organisational structure made data-sharing difficult, if not impossible;

Achieving Business Success with GIS   Bruce Douglas
© 2008 John Wiley & Sons, Ltd   ISBN: 978-0-470-72724-9

- all of the costs were not taken into account and the project cost and time substantially 'blew-out';
- very few of the benefits were able to be realised and those that were able to be realised were a lot less than estimated;
- staff opposition to the new system was insurmountable.

In many organisations, IT (and GIS) budgets are shrinking while expectations from users continue to expand. This often results in increasing pressure for projects to become more focused on targeting business 'hot spots' for technology initiatives as well as focusing on proving (and documenting) any business benefits that have been achieved. Therefore in order for this to occur, it is important to fully understand and document:

- the costs over the 5-year period, including the 'total cost of ownership' of the system;
- the estimated tangible and intangible benefits (including operational and strategic benefits) over that period; and
- the estimated NPV of the investment and the ROI.

This chapter outlines the process of undertaking a Cost/Benefit Analysis for consideration by senior management. But first, before any words are written on this subject – *beware*, all costs and benefits gathered and used in any Cost/Benefit Analysis should be done on the basis of 'full disclosure'. That is, costs should not be 'glossed over' and benefits should not be 'painted in the best light'. To do so will only 'set the project up' for failure in the long term.

It is highly likely, in fact it is almost guaranteed, that there will be cost over-runs. It is also highly likely that the benefits that are described will not be able to be realised in full. For example, if a benefit is based on a 10% efficiency gain for 50 staff, this will equate to reducing staff costs by five people. That is, there will be an expectation by senior management that the total staff budget will reduce by the equivalent of five persons' salaries (effectively five staff leaving the organisation). However, if the expectation is that staff will not be made redundant but re-deployed to another role in the organisation (and therefore their salaries will remain) then this should be stated as such.

Therefore, it is highly recommended that any Cost/Benefit Analysis be 'fiscally very prudent' and be 'told as it is'. This will not only gain the respect of senior management (they are not fools after all), but will set the scene for dealing with future project over-runs, just in case they happen (and they will happen).

Again, as in the case of defining the GIS Strategy, there is no such thing as a generic Cost/Benefit Analysis or Business Case. Just as every organisation is different, so too is every GIS Cost/Benefit Analysis different.

At this stage, it is also worthwhile noting (as discussed in previous chapters) that while a system management methodology can provide a useful set of 'cook-book' tools, these methodologies are only moderately useful to this process, as spatial environments often have different sets of issues to be addressed.

When developing a Cost/Benefit Analysis and Business Case, it is true to state that:

- the costs are usually the 'easy bits' – they can be gathered/estimated with some degree of methodological process – and if they cannot be 'pinned down', apply a contingency of at least 50% to the 'best estimate';

• the benefits, particularly the intangible and strategic benefits, are usually the most difficult to gather and to document in a form that is credible (i.e. able to be believed) by senior management.

The costs and benefits discussed below are done so in the context of a generic organisation even though they have been abstracted from actual projects.

9.1   BROAD COSTS

The costs associated with the implementation/expansion of a GIS environment can be categorised as encompassing the following:

1. Organisational Costs
2. Software Costs
3. Hardware and Infrastructure Costs
4. Database Development Costs
5. Data Capture and Conversion Costs
6. Other Costs

In the doctrine of 'full disclosure', not 'glossing over' costs, being 'fiscally very prudent' and 'telling it like it is', the costs described below should be real. That is, the costs should be based on the assumption that if the Business Case is approved and the budget is granted, then these costs should be the absolute *maximum* that should be spent on the project.

The corollary is that it is just as difficult to request (say) $850 000 as it is to re-quest $650 000, so if the project is really going to cost $850 000, then why ask for only $650 000 knowing that you will only have to go back again and ask for another $200 000 or cut the project down to fit the budget.

The following costs are extracted from particular case studies and relate to the imple-mentation of a corporate SI environment.

ORGANISATIONAL COSTS

Most GIS projects will require a dedicated resource (or resources) in order for the GIS environment to be set up, the data to be organised and all the 'corporate' tasks necessary to make the GIS operational to be undertaken. It is usual for a GIS 'Work Team' to be appointed to undertake these tasks, the size of the team being dependent on the amount of work required to be undertaken.

And when considering the formation of a new GIS team, the organisational location of this team is most important. Results from the GIS/Spatial Best Practice Surveys indicate that for all industry sectors, except for municipal councils, the most common reporting focus of the GIS staff was to a manager with a relevant business focus, that is, the Asset Manager, the Director of Planning, the Network Manager, etc.

However, for some organisations, the GIS staff report to a manager with an information technology focus – in this regard municipal councils are unique and this divergence from the most commonly used practice by other industry sectors is likely due to the more generic or widespread application of GIS technology across the wide range of business functions performed by municipal councils.

While these reporting arrangements can offer a practical management model for organisations with diverse business responsibilities, it is generally considered that such arrangements make it more difficult to implement an effective Purchaser/Provider business model[1] for GIS technology service delivery.

In this example, a 'GIS Work Team' will need to be formed in order to undertake the 'corporate' tasks outlined in the GIS Strategy. This group should organisationally reside as part of a business unit (e.g. in the Planning and Development Team). In this example, a GIS Work Team of two people, including 30% on-costs,[2] might be required with a cost as follows:

- one GIS Administrator at $55 000 per annum; and
- one GIS Operator at $45 000 per annum.

The costs for the GIS Work Team would therefore be $130 000 per annum.[3]

It was not expected that there would be any other organisational costs incurred 'over and above' the normal day-to-day activities of other staff surrounding the GIS team. While some types of data being supplied to, and received from, the GIS team would change, it was not envisaged that there would be any additional time (and therefore cost) imput because of this. Therefore the total organisational costs were estimated at $130 000 per annum.

In order to ensure that all GIS staff are trained to the appropriate level required, commensurate with their discipline and their expected usage of various spatial applications, it is envisaged that a considerable amount of training will be required. Industry experience is that training is required and that it must be in the context of the work being undertaken by staff.

Therefore in order to ensure that the skilling is retained and is context sensitive to the various roles within host organisation, it is important that a set of courses be arranged using examples targeted to the GIS team members undergoing the training. A budget of $20,000 should be allocated for the provision of a number of training courses, and (optionally) making use of "train the trainer" courses to develop these training skills in-house.

SOFTWARE COSTS

While contemporary industry trends are to 'buy not build', there are some merits in developing software which uniquely meets the needs of the business. However, the result of this build mentality is usually that the product becomes an orphan, totally separated from the core software from which it was derived and therefore unable to be brought forward to the next software version without a total re-build often at a considerable cost. Nevertheless, the total cost of ownership should be the determinant of this decision, particularly after including the cost of software built to align totally to the business process versus one that is not.

This case study required that a Corporate GIS be developed, based on an Oracle Spatial database with a number of GIS 'seats' being acquired in addition to the two licenses for the GIS team. A breakdown of the expected costs that would be incurred for software follows, ending with a summary.

---

[1] Note that Purchaser/Provider models do not necessarily suggest that a regime of internal charging be applied.
[2] 'On-costs' typically include all the other costs that are required for an employee to work, e.g. phone, desk, chair, floor space, lighting, superannuation, leave costs, etc. In most government organisations, a 30% 'on-cost' loading is considered typical, but on a mine site with fly-in fly-out staff (for example) the on-costs could be as high as 150%. Note that this will vary from region to region and from industry to industry. This percentage rate should be confirmed with the organisation's Finance Manager.
[3] 1 at $55 000 + 1 at $45 000 = $100 000 + 30% on-costs = $130 000.

GIS Software

The GIS Strategy indicated that the GIS software requirements would consist of (for this example):

- GIS Edit licenses (3–5 licenses);
- GIS Viewing (intranet application for all staff);
- GIS PDA/GPS licenses (1–3 licenses); and
- an image processing software application if required.

Indicative costs provided by the GIS suppliers indicated that the GIS software should cost approximately $70 000 and that a further $10 000–15 000 should be allowed for image processing software (if required).

Tip

In trying to gain an understanding of the GIS software costs, it can be useful to discuss the project with a couple of vendors and request 'indicative costs' which may be appropriate should they be required to implement this system.

But Beware – by discussing the project with vendors at this stage, it is possible that they come to think that they have already won the business. Given that this project may (should) go to an open tender if approved, extreme care should be taken when dealing with vendors to ensure that expectations are not being set (with the vendor) and that probity is not un-knowingly being prematurely breeched. Therefore, it is always best to conduct this in writing (email) with several vendors and in such a manner that each is aware that other vendors are also being consulted.

Given that it is expected that further costs may be uncovered when a firm quotation is provided, an allowance of approximately $100 000 should be made to provide this GIS environment which can support multiple Edit and View users across an intranet, including minor customisation, implementation, training and setting-to-work of the system.

Note that while the maintenance cost of the GIS software is typically 10%, some vendors have higher maintenance costs for specific items in the software suite, depending on the specific GIS software selected. Therefore a higher than typical maintenance cost of 13%–15% p.a. (or even 20% in some cases) should be allocated to allow for this contingency.

The additional cost of the interfaces and intranet viewing applications required for the Corporate GIS are discussed below.

Corporate Database

The development of a corporate database will require the development of a Corporate data model and data dictionary, creation of the physical data model and loading of data into this database. Since the client already had Oracle Enterprise licenses (server based) for Oracle, discussions with Oracle suggested that the spatial extensions to the database would cost of the order of $58 000 with an annual maintenance of $12 000.

While there may be some issues which could result in these prices being discounted, it would be prudent to leave these costs at the above amount and increase them slightly to cover unforeseen issues. Therefore, a cost for the corporate database licensing of $70 000 should be allowed.

The costs for development of the database and loading of the data are included under 'Database Development Costs' below.

Interfaces

The GIS Strategy indicated that there may be a need to have interfaces to and from the Financial system, subject to further investigation being undertaken during the design phase. The Financial system also contained considerable data relating to assets and there was a need to use this asset data (from the Financial system) in the GIS and (in some limited cases) vice versa.

While it was probable that the need for an interface with the Financial system could be accommodated by loading this financial data into an external database at periodic intervals, this had yet to be decided. Therefore, while it is expected that this requirement will not incur any costs apart from some internal scripting, an amount of $10 000 is allowed as a contingency should development be required.

Lease Management

The strategy called for a Lease Management capability to be acquired and sourced from third-party application developers (several had been inspected prior to this stage). While the scope of the Lease Management application had yet to be fully developed, it was expected that a budget of approximately $20 000 should be appropriate for this purpose.

PDA Software – Asset Checking

The GIS Strategy indicated that there was a need to have a PDA application to allow staff to input an Asset Number (or swipe a barcode) and have returned the Asset History which had been loaded onto the PDA from the corporate database using data previously loaded from the Financial system. This should also include the ability to update the asset details, record faults which might drive changes in maintenance programmes and to collect details of new assets.

It is expected that the software module for PDAs would be able to be installed on the PDAs with the GPS cards installed. An allowance of $8 000 should be allocated for the purchase/development of specific software for three PDAs to allow specific field operations to be expedited. Note that this does not include the PDA hardware costs and this is in addition to the PDA software costs included under 'GIS software' above.

Intranet Viewing

The GIS Strategy outlined the need to have a 'view' capability of data across the organisation's intranet, based upon an Internet Explorer web-browser. The implementation of this requirement will be dependent on the particular GIS solution selected and is expected to be included in the pricing under 'GIS Software' above. No specific intranet business applications were required other than a general view requirement, typically an 'out-of-the-box' web-mapping capability for most GIS products.

Metadata Manager

A Metadata management tool will be required to provide a capability to input, manage and retrieve metadata from the corporate database, particularly for the intranet viewing application. The research suggested that while a couple of tools were available, this capability would probably need to be developed and an allowance of $20 000 should be budgeted to create this tool, subject to the capabilities of the GIS solution (yet to be selected).

Summary – Software Costs

The software costs as discussed above can be summarised in the following table.

Software	Estimated Purchase Cost	Estimated Maintenance Cost p.a.
GIS Software	$100 000	$14 500
Corporate Database	$ 70 000	$12 000
Interfaces	$ 10 000	
Lease Management	$ 20 000	
PDA Software	$ 8 000	
Internet Viewing Capability	Inc in GIS	
Metadata Manager	$ 20 000	$ 2 000
Total	**$228 000**	**$28 500**

HARDWARE AND IT INFRASTRUCTURE COSTS

The GIS Strategy required that a Corporate GIS be developed based on the current level of IT infrastructure, which was generally of high quality with excellent networking capabilities. The only items required are shown below.

PCs

While an additional three PCs will be required for the GIS team (two staff plus a development PC), two additional PCs will also be needed for the two emergency control rooms. Therefore, a total of five high-end PCs have been included for budgeting purposes at $6500 each, totalling $32 500.

Additional Server

It was expected that an additional server would be required to support the GIS implementation and to provide the level of redundancy required. Therefore a cost of $14 500 should be allocated for this purpose, to be located in the existing server room.

Printers/Plotters

The widespread use of GIS often results in additional staff producing more printing at a larger (than usual) format.[4] While this may result in a demand for more printing devices, the costs of these printers have not been included here, except for those required by the GIS team – an additional plotter at an estimated cost of $10 000.

PDAs

The cost for at least three PDAs with GPS should be included in this implementation even though it is expected that this number of devices will grow substantially over coming years. At a unit cost of approximately $1500 for a GPS-enabled PDA, a cost of $4500 should be allowed (note that the PDA software development costs are given in the table above for software costs).

Summary – Hardware and IT Infrastructure Costs

The following table outlines the estimated costs for the hardware/infrastructure for the proposed Corporate GIS.

Hardware & IT Infrastructure	Estimated Purchase Cost	Estimated Maintenance Cost p.a.
PCs	$32 000	Internal
Additional Server	$14 500	Internal
Printers/Plotters	$10 000	$1000*
PDAs	$ 4500	$ 450*
Total	**$61 000**	**$ 1450**

*Note: Maintenance has been estimated at 10% of purchase price for these items.

DATABASE DEVELOPMENT COSTS

The GIS Strategy required that a Corporate GIS be developed based on a corporate database environment. Therefore a data dictionary and data model would need to be developed as part of the overall detailed design of this environment, particularly to ensure that there is a seamless link to the Financial and Asset Management corporate databases/systems within the organisation. In addition, the data currently resident in the multiple Access databases should be rationalized, with necessary data being migrated to the corporate database and the Access databases being subsequently de-commissioned. When complete, the data from each of these above systems should be able to be accessed by the GIS, using a unique asset / equipment identifier.

It would be expected that an analyst familiar with data structures, Oracle Spatial and the selected GIS would be required to undertake these tasks in conjunction with the organisation's IT staff. While it is difficult to be precise about the cost of these tasks,

---

[4] Printing and Plotting from GIS frequently requires the use of larger than normal paper, in some cases up to A0 size, as well as the ability to plot in colour. Therefore while A3 and A4 printers are commonly used, an A0 plotter can be very useful for specific tasks.

particularly prior to the selection of the GIS, a budget of $50 000–80 000 should be allowed to:

- create the existing corporate data dictionary and ensure compatibility with the Financial and Asset Management systems and the existing Access databases;
- develop a data model; and
- design and build the Oracle Spatial database ready for populating.

Therefore a budget of $80 000 should be allocated for this task, with an additional $5000 each year for minor additional development tasks.

Note, however, that while this case study focused on the organisation's legacy data, other data sources are often available, such as that available from vendors who sell ready-to-use data (e.g. topography, imagery, street-mapping, competitive intelligence, industry public-domain data) in pre-cleaned proprietary software formats. While these vendors can be a good resource for GIS data, the use of their services is often not cheap and should be costed into the project.

DATA CAPTURE/CONVERSION COSTS

For the Corporate GIS to be implemented correctly, a number of datasets will need to be captured/converted from hard-copy drawings and from the field. It should be noted that all GIS environments need a large amount of 'base' data in order to undertake the tasks often required of the GIS. Should this data not be available, either through cost considerations or lack of availability of the data, the business functions (which use this data) will most likely not be able to be undertaken to any extent.

The GIS Strategy outlined that the following data was required to be captured and/or converted to enable the GIS to become operational. The costs for this data are detailed below.

Capture of Aerial Photography

The GIS will require the capture of all major infrastructure from aerial photography or ground survey. While the latter (ground survey) would provide a higher level of accuracy, it was envisaged that the cost would be prohibitive and would take an inordinate amount of time.

Digital orthophotography of the area in question and environs was determined as being the most suitable to meet the intended business functions of the GIS. This photography should be undertaken at a photo scale of approximately 1 : 5000 to achieve an on-ground accuracy of 0.1–0.25 m. This will require some on-ground survey work, including placement of control marks for the photography, post-photography ground verification and augmentation.

Discussions with several aerial surveying companies indicated that the cost of undertaking this photography, digitising the data, undertaking field control and validation and converting the data to a form suitable for the corporate GIS (using software yet to be purchased) would be at least $60 000 and could be as high as $100 000.

In addition, there would be some time required by client staff to manage this process, to receive and check the data and to load the data into the GIS. It is estimated that over the duration of this project, a total of approximately 75 person-days of effort will be required by several contract and GIS staff. A budget of $20 000[5] should be allocated for this process. Therefore, the total cost to be budgeted should be $120 000.

Should the data not be captured from aerial photography, most of the above information necessary to populate the GIS will need to be captured by digitising old drawings, augmented by traditional ground processes to capture the data (mostly by engineering and surveying staff). While it is difficult to estimate this alternative cost, this will take at least 6 months for a two-person digitising team augmented by on-ground survey work for several months – to give a cost which could be as high as $150 000.

Interpretation and Digitising of Underground Infrastructure Data

Once the digital orthophotos have been captured, there will need to be some GIS work undertaken to manually locate and digitise underground infrastructure between the locations of gully-pits, man-holes, etc., captured from the digital orthophotos. This will also require some digitising from a number of the old drawings as well as some ground truthing of these underground services over time.

It is expected that a GIS operator will need approximately 50 days to undertake this work at a cost of approximately $13 300.[6]

Financial/Asset Data Review/Update/Correlation

The data currently available in the Financial and Asset systems will need to be reviewed, updated and correlated. This should include the review of existing databases and processes in order to improve the level of data integrity in the Financial system, and subsequently in the data abstracted from that system to the corporate database. While this could be a very large task (and could be a major project in itself), an analyst for approximately 3 person-months would be necessary to review the scope of work required, to undertake some preliminary work and to further define this project scope. As such, a cost of approximately $31 900[7] is allowed for an analyst for 3 person-months.

Note that this cost is only for an analyst to scope the much larger task and does not include doing the major project – such a major project is considered to be a separate (non-GIS) project which should be undertaken in parallel to the GIS implementation.

Collection and Conversion of Data from External Agencies

A small amount of data will be required to be collected from the local Councils and the state Lands Department, etc. While most of this data will be available in digital form, it is expected that there may be some amount of time required to identify these datasets, to arrange for the collection, to input the data, convert the data to the format of the client systems (yet to be defined) and then to massage the data into a form that is useful to the organisation.

It is expected that a GIS Officer will need approximately 50 days to undertake this work at a cost of approximately $13 300.[8]

---

[5] GIS Officer for 75 days = ($45 000 + 30% on-costs) / 220 × 75 = $19 950.
[6] GIS Operator for 50 days = ($45 000 + 30% on-costs) / 220 × 50 = $13 295.
[7] Data Analyst for 3 months = 90 days = ($60 000 + 30% on-costs) / 220 × 90 = $31 909.
[8] GIS Officer for 50 days = ($45 000 + 30% on-costs) / 220 × 50 = $13 295.

Loading Data into Corporate Database

The GIS Strategy was based upon the deployment of data in a corporate database using Oracle Spatial for all data which will be accessed by the GIS.

The users of this data will be able to access this information as:

- GIS Edit users who will use GIS software to read and write to the GIS and corporate database; and/or
- GIS View users who would use an Internet Explorer application to read from the GIS and corporate database to present integrated data.

The strategy outlined the costs involved in developing a data dictionary, data model and the build of the database to a state which would be ready for populating with data. The work involved in this process would include the data collection, manipulation, sorting and loading into a corporate database as well as ensuring that any interfacing to other corporate databases (e.g. Financial system) is structured correctly. It is expected that a GIS Administrator will need 2 months at a cost of approximately $19 500[9] to undertake this task.

Summary – Data Loading, Capture, Conversion and Integration

In summary the following table outlines the estimated costs to collect, convert, organise and manage the spatial and textual data for use in a Corporate GIS and database environment.

However, in reviewing this, it should be recognised that a considerable amount of this work will be required to be undertaken by in-house staff. This may require existing staff to be taken off-line to undertake these duties (and perhaps being back-filled by contract staff) or by hiring contract staff to undertake some of this work. Regardless of the method used, this table outlines the costs expected to be incurred in this process.

Note also that it is not envisaged that new GIS staff will be able to undertake a lot of this work for the first 3–6 months as well as his/her normal GIS Operator duties, such as overseeing the implementing of the GIS system, liaising with IT and CAD, undertaking training, etc. While a number of these tasks can be undertaken in parallel, it is estimated that the total elapsed time required to undertake the data components necessary for implementation of the Corporate GIS would be 6–12 months.

Data	Estimated Cost	Resources Required (person-months)
Capture of Aerial Photography	$120 000*	GIS Officer – 2
Interpretation/Digitising of Underground Infrastructure Data	$13 300	GIS Officer – 3
Financial/Asset Data Review/Update/ Correlation	$31 900	Data Analyst – 3
Collection & Conversion of Data from External Agencies	$13 300	GIS Officer – 2
Set-up/Loading Data into Corporate Database	$19 500	Data Analyst – 2
Total	**$198 000**	**12**

*Note: Plus an additional cost of $25 000 for the re-capture of the digital orthophotos every 3–5 years.

[9] GIS Administrator for 2 months = ($55 000 + 30% on-costs) / 220 × 60 = $19 500.

OTHER COSTS

In addition to the costs discussed above, there may also be further costs required to complete this project, e.g.:

- the tender evaluation may require travel to undertake reference site checks;
- the project team may retain the services of a consultant to assist with some components of the above processes; and
- there may be additional, as yet unspecified, costs.

However, in this case study it was considered that there would be no additional costs.

SUMMARY – COSTS

The costs for the development and implementation of a Corporate GIS, including the development of a corporate database, can be summarised in the following table.

Cost	Initial Costs	On-going Costs p.a.
Organisational	$150 000	
Software	$228 000	$28 500
Hardware & IT Infrastructure	$ 61 000	$ 1450
Database Development	$ 80 000	$ 5000
Data Capture/Conversion Costs	$198 000	
Other Costs	nil	
Total	**$717 000**	**$34 950**

As such, it would be prudent to add at least a 10% contingency and consider that the budget requirement should be rounded up to $850 000.

In summary it should be noted that:

- the total budget for the 'GIS environment' is $850 000; but
- the original estimated cost of just the GIS software was $70 000.

This is particularly important – industry experience is that the total cost for a GIS environment is usually between 10 and 20 times the cost of just the GIS software.

Fact (for this case)

GIS Software = $70 000
Total GIS Environment = $850 000

The 'total cost of GIS implementation' is typically 10–20 times the cost of the GIS software.

Unfortunately when GIS is proposed in a number of organisations, there is a tendency to focus on the GIS software as the 'total cost' – much to the peril of the person proposing this tactic. Should this occur, and senior management approve a budget of just $70 000 to purchase the GIS, imagine their concern when they realise that the budget has climbed to over three-quarters of a million dollars!

Therefore, one of the major objectives of undertaking the Cost/Benefit Analysis is to ensure that all of the costs are identified and documented in some detail before the project commences (when there is still the luxury of doing something about the costs) rather than after the costs are incurred.

9.2   BROAD BENEFITS

The implementation of GIS technology into an organisation can often provide substantial benefits, both tangible and intangible. These benefits can also be categorised as operational (i.e. benefits that are realisable and tangible) and strategic (often the intangible 'value add' that comes from being able to 'do more with less').

However, because GIS is 'data-centric', the implementation of GIS usually requires that a large amount of data is captured, converted and cleansed before it can produce any benefits. Therefore this takes some time and often comes at some (often considerable) cost.

Nevertheless, if the GIS is implemented in a comprehensive and methodical manner, substantial business benefits can be gained, most often categorised as tangible and intangible benefits.

Tangible benefits are those which can be realised as specific monetary amounts accrued to the organisation as a result of specific activities, such as:

- saving or avoiding future costs (that may occur if GIS was not implemented);
- better use of resources (IT and human);
- improving efficiency, i.e. doing things faster and easier;
- increasing income;
- reducing cost/waste.

Intangible benefits, on the other hand, are often very difficult to quantify and are generally not able to be costed, except perhaps in very broad terms. Nevertheless, intangible benefits are valid and can often make significant improvements to the business of the organisation. Intangible benefits may be able to be derived from:

- being able to do better planning;
- having better environmental management;
- having better management of assets;
- having better compliance;
- avoiding risk;
- improving the 'attractiveness' of the organisation's business.

The following benefits are examples of typical benefits that have been abstracted from actual projects and presented in a generic form.

Note again that in the doctrine of 'full disclosure', not 'embellishing' benefits, being 'fiscally very prudent' and 'telling it like it is', the benefits should always be real – and they should be based on the assumption that if the Business Case is approved and the budget is granted, then these benefits should be able to be *realised*, i.e. the monetary value of the benefit promised should be able to be re-couped by the organisation.

BENEFITS – TANGIBLE

The tangible benefits which may be able to be accrued from the implementation of a GIS Strategy would include the following – note that the following categories are examples from previous case studies and would not all apply to all given benefits determinations.

Fire Risk Mitigation

On average, there are four bushfires each year in the region which is operated by the subject company. A recent fire resulted in five power poles being destroyed and power being interrupted for 8 hours. This resulted in a lost production cost of approximately $50 000. On average there are 4–6 fires per year and if this happens four times per year a loss of $200 000 could be incurred even though it is estimated that there could be a substantially higher exposure, perhaps as high as a million dollars if a major fire interrupted power and water production for a prolonged period.

The implementation of a GIS will enable better knowledge of assets (and the condition of those assets), which should enable a faster response with less down-time, should those assets be damaged. In addition, knowledge of the condition of the assets will enable assemblage of materials for quick repair, should that become necessary. Therefore, while a GIS will not avoid incurring damage to power utilities, it will severely lessen the impact that the risk of a fire has on production. As such, it is estimated that at least half of these costs can be avoided when the GIS is operational. A conservative estimate of the benefits is therefore half of the above $200 000 p.a., i.e. $100 000 p.a.

Flood Risk Mitigation

The recent rain and subsequent minor floods have resulted in damage to utilities with resultant loss of production. Because the roads to a number of utility services are in poor condition, and because there is little local knowledge of the surrounding terrain, considerable time is lost determining the cause of the problems before they are able to be fixed. While these floods may be considered as an isolated incident in drought conditions, in normal years there is generally several floods each year, each with consequent ongoing problems of determining access. It is estimated that at least 2 days are lost by six staff (three crews of two each) due to lack of information to enable easy access to utility services.

This amounts to costs of $9500 p.a.,[10] which could be avoided if a GIS were implemented and if correct information was known about the location of access tracks as well as swamps and gullies.

Better Environmental Management

Discussions with stakeholders indicated that there are several minor spillages from the retention dams each year which could be helped by better mapping of tailings, thereby providing more surety that any tailings do not result in contaminations which may cause breaches of environmental regulations with possible attending fines.

---

[10] Two days for six staff for three times p.a. = 36 days p.a. = $9500 for a salary of $45 000 + 30% on-costs.

While it is difficult to estimate the benefits that could result from improved management of tailings, they could be significant. If a GIS were implemented so that better mapping could be provided and correlated with environmentally sensitive areas, it is estimated that at least $50 000 p.a. could be avoided in Environmental Protection Agency fines – noting that this level of fines has occurred several times in recent years. In addition, stakeholders considered that such an information repository could also be instrumental in providing better environmental outcomes.

Better Lease Management

It is known that there could be substantial benefits from having improved management of leases, including:

- a considerable amount of time is lost by staff searching for information relating to leases;
- a considerable amount of production time could be avoid being 'bottle-necked' or lost if better management practices could be put in place for the management of the many forms and notices relevant to leases;
- a considerable amount of bond moneys may be able to be avoided if better management of leases could be put in place; and
- a considerable amount of money could be saved from claims if more information were known about leases and not based on individual's memory.

While these costs are not known or able to be calculated with any precision, it is known that they are substantial. Stakeholders considered that it would not be unrealistic for there to be benefits of at least $100 000 p.a. from having better lease management practices in place, based upon an operational GIS.

Planning Enquiries and Permit Applications

Approximately 2 person-days per week (on average) are spent by the Planning staff searching for information relating to land, leases, licenses, services, etc. Often this work-load is spread over a 1–2-week period with time being spent waiting for staff from other areas (e.g. drafting, engineering, etc.) to advise issues and conditions relating to the subject enquiry.

While this process unnecessarily wastes the time of Planning staff, it also creates an unprofessional image when dealing with potential customers. This is further exacerbated by subsequent changes, options, etc. after the original information is collated and presented. The information most often required relates to conditions of leases, licenses, encumbrances, services, property history (e.g. contaminates, etc.) in order for the potential customer to make an informed decision.

It is estimated that most of these 2 person-days per week could be saved if a GIS were implemented and if information were readily available, resulting in a benefit which could be as high as $43 000 p.a.[11]

---

[11] 1.5 person-days per week × 52 weeks = 1.5 × 52 × $60 000 / 220 = $21 200 p.a.

Unidentified Works

A number of (mainly engineering) projects and incidents have been undertaken which could have potentially been avoided (and therefore produced cost savings) if information about the assets was available prior to commencement of the project/incident and if that information was correct and up-to-date.

Stakeholders discussed costs in excess of $250 000 p.a. which could have been avoided if the correct information was known before commencement of these projects. While these particular projects were 'one-offs' and will not occur again, they are typical of other projects which will occur and will incur similar problems with regard to information about assets. Therefore, it is apparent with the continuing number of incidents such as these that there is a high probability that there will be other ongoing benefits that could be obtained from having a good quality database about assets, particularly one that is based upon a GIS.

As such, while it is not realistic to allocate the $250 000 as a benefit, at least 30% of this could be allocated as a tangible benefit. However, as a conservative estimate, only 15% (i.e., $37 500 p.a.) is taken as a tangible benefit if such unidentified works could have access to good quality data so that the cost of remediation could be minimised (or avoided) in the future.

More Focused ad hoc Mapping

Since the disbanding of the Cartographic team, Surveyors have periodically been required to produce a range of ad hoc maps for a variety of purposes using the time spent not being devoted to their surveying tasks. Stakeholders agreed that there is an ongoing need to provide a capability to deliver map information to meet a number of the organisation's business needs. The GIS Work Group resulting from this Corporate GIS will be able to provide this facility in a consistent and managed manner, thus freeing up the Surveyors to focus on their core tasks. It is estimated that at least 5 days per month of a Surveyor's time could be saved if this work were undertaken by the GIS group, giving a benefit of $27 300.[12]

More Effective Site Investigation

When the current project data holdings of the group (estimated at hundreds of drawings and approximately 300 CDs) are reviewed, catalogued and loaded into the GIS, it is estimated that there could be substantial benefits (derived as costs avoided) from:

- avoiding site investigation in either the wrong place or in places which have already been investigated;
- not knowing whether all of the data is at hand;
- avoiding lost opportunity costs; and
- avoiding potential litigation costs.

While there are no statistics which can be used to indicate the size of these benefits, it is conservatively estimated that this avoided cost could be tens of thousands of dollars (and

---

[12] Two days per week = 104 days p.a. = $43 000 p.a. for a salary of $70 000 + 30% on-costs.

possibly millions of dollars in terms of opportunity cost) each year if this data issue is not addressed.

Even if a conservative estimate were taken of $50 000 p.a., this benefit could be substantial.

Summary – Tangible Benefits

The following table summarises the estimated tangible benefits which may result from the implementation of a Corporate GIS Strategy.

Tangible Benefits	Estimated Benefit p.a.
Fire Risk Mitigation	$100 000
Flood Risk Mitigation	$9500
Better Environmental Management	$50 000
Better Lease Management	$100 000
Planning Enquiries and Permit Applications	$43 000
Unidentified Works	$37 500
More Focused ad hoc Mapping	$27 300
More Effective Site Investigation	$50 000
Total	**$ 417 300**

BENEFITS – INTANGIBLE

In addition to the tangible benefits discussed above, a number of benefits cannot be costed and are therefore intangible in nature. Nevertheless, benefits such as these are generally able to be achieved from the implementation of this GIS Strategy.

The following intangible benefits are presented, recognising that while most would agree that they are significant and may provide a substantial improvement in the delivery of many corporate initiatives, they are intangible and are therefore unable to be estimated, except in very broad terms.

Less Production Risk

While the examples of the bushfires and floods cited above have specific tangible benefits if avoided, a better understanding of the risk of such an event should provide more surety about particular risk avoidance strategies, hopefully averting that which could otherwise become a major incident, thereby leading to higher levels of production loss. The GIS Strategy will provide better information, and easy access to it, which should provide a higher level of understanding of a risk, and the need to mitigate that risk, to production.

Better Safety and Health

Staff safety operating in remote locations is important, particularly to ensure that staff can be located easily if incapacitated. Unfortunately with no aerial photography being available, it is difficult to know the location and condition of roads and tracks, with most staff relying on local knowledge for information about local conditions. The use of aerial photography on the GIS, in conjunction with GPS technology on PDAs, can provide immediate and relevant information to communicate to searchers for rescue. However, this benefit is intangible in nature and cannot be costed.

More Effective Emergency Response

The ability to quickly assess and interpret information about the regional environment in an emergency situation will enable more effective use of resources in combating an emergency, including:

- access for fire-fighting equipment and planning of fire breaks;
- better understanding of egress routes for protection of emergency workers; and
- better understanding of water points such as creeks for fire planning.

Fast access to this information is vital to provide a timely response to an incident and to minimise risk, potentially saving lives and reducing the time and cost of response and remediation.

More Effective Lease Management

In order that each lease is run as a profitable enterprise, there is a requirement to know more about improvements (windmills, yards, fences, gates, etc.), work to be undertaken, pasture type/condition and stocking and carrying capacity. The information provided by the Corporate GIS will allow more effective management of these leases so that every effort can be made to maximise the profit potential from these enterprises. While this benefit is also included with the tangible benefits, there are also a number of intangible benefits in this category that justifies that it be in both categories. However, care needs to be taken to ensure that 'double counting' is not occurring.

Plume Modelling Capability

The implementation of a Corporate GIS may provide the capability to model the plumes of toxic gases should the occasion arise. While this will require the addition of specialised software to the GIS, the planned GIS will provide the basis for this capability should it be required. Such plume modelling will provide more effective emergency response as well as a more responsible and responsive environmental management capability.

Better Environmental Management

Better information about the environment will provide the ability for staff to provide a higher level of environmental management, thus avoiding costly fines and remediation should an incident occur, as has happened several times in the past several years. Again, while this benefit is also included with the tangible benefits, there are also a number of intangible benefits in this category that justifies that it be in both categories.

INTANGIBLE BENEFITS – OTHER FORMS OF COSTING

Although intangible benefits are subjective and unable to be costed, experience from large high-risk companies, such as oil exploration, is that one often has to attempt to quantify the unquantifiable with dollars. In such a situation a simple formula may be useful:[13]

$$\text{Intangible Benefit} = \text{Cost of Risk Exposure} \times \text{Probability}$$

where: Cost of Risk Exposure is the maximum cost of risk exposure or the value of potential business opportunity; and Probability means the probability of this happening if the proposed GIS project is not going ahead, optionally multiplied by the % role of GIS in identifying the risk/opportunity.

For example, if there is a missed opportunity of not using GIS for a project with a potential value of say $100 million, and the probability of this happening is 1%, the intangible 'benefit' of the GIS would be $100 million $\times$ 1% = $1 million. However, if one then went further and said that two-thirds of the benefit may be due to other factors, such as the knowledge of the engineers/geologists on the project, and that one-third is due to the GIS providing the data and framework to realise the benefit, then it can be argued that the benefit of the GIS is one-third of $1 million, which is approximately $300 000. In the business of a high-risk speculative venture, this is often a useful and practical estimate-of-benefit calculation.

9.3   BROAD COST/BENEFIT SUMMARY

The results of the broad costs and benefits as discussed above can be summarised in the following table and used to determine the Net Present Value (NPV) for implementing the GIS Strategy.

The Net Present Value (NPV) is the 'value' of this cost / benefit model 'brought forward' to today's dollars. That is, the table shows a Net Benefit of −$467 541 in Year 1, a Net Benefit of $57 591 in Year 2, a further Net Benefit of $252 350 in Year 3, and so on. If all of these Net Benefits are brought back to the 'Year 0', taking into account the interest rates which would apply to these amounts and summed, the NPV can be calculated as follows:

- NPV at 8% = $160 680
- NPV at 10% = $129 090

If there is a positive NPV, then the project is worth investing in (financially). Conversely, if the NPV is negative, then this project may not be considered a financially good option.

Note, however, that the intangible benefits are not included in these calculations, and when included they may considerably outweigh any concerns resulting from a poor cost/ benefit calculation.

---

[13] Thierry Gregorius, Head Data Management and Geomatics, Global Exploration, Shell International Exploration and Production, The Netherlands.

	Unit Cost	Maintenance Costs	Note	Year 1	Year 2	Year 3	Year 4	Year 5
Personnel								
GIS Administrator × 1	$55 000		1	$71 500	$71 500	$71 500	$71 500	$71 500
GIS Operators × 2	$45 000		1	$29 250	$58 500	$58 500	$58 500	$58 500
Training	$20 000			$20 000				
Sub-total Personnel	$120 000			$120 750	$130 000	$130 000	$130 000	$130 000
Software								
GIS Software	$100 000	$14 500	2	$100 000	$14 500	$14 500	$14 500	$14 500
Corporate Database	$70 000	$12 000		$70 000	$12 000	$12 000	$12 000	$12 000
Interfaces	$10 000			$10 000				
LeaseManagement	$20 000		2		$20 000	$0	$0	$0
PDA Software	$8000				$8000			
Internet Viewing Capability	Inc in GIS		2	Inc in GIS				
Metadata Manager	$20 000	$2000	2	$20 000	$2000	$2000	$2000	$2000
Sub-total Software	$228 000	$28 500		$200 000	$56 500	$28 500	$28 500	$28 500
Hardware & IT Infrastructure								
PCs	$32 000			$32 000				
Additional Server	$14 500			$14 500				
Printers/Plotter	$10 000	$1000		$10 000	$1000	$1000	$1000	$1000
PDAs	$4500	$450			$4500	$450	$450	$450
Sub-total Hardware and IT	$61 000			$56 500	$5500	$1450	$1450	$1450
Database Development								
Database Development	$80 000			$80 000		$5000	$5000	$5000
Sub-total Database Development	$80 000			$80 000		$5000	$5000	$5000
Data Capture/Conversion Costs								
Capture of Aerial Photography	$120 000			$90 000	$30 000			$25 000
Interpretation and Digitising of Underground Data	$13 300			$13 300				
Financial/Asset Data Review & Update	$31 900			$31 900				

Collection & Conversion of Data from External Agencies	$13 300		$13 300				
Set-up/Loading Data into Corporate Database	$19 500		$19 500				
Sub-total Data Capture/Conversion	$198 000		$168 000	$30 000	$0	$0	$25 000
Total Costs (ex org)	**$567 000**		**$625 250**	**$222 000**	**$164 950**	**$159 950**	**$189 950**
Cumulative Costs			$625 250	$847 250	$1 012 200	$1 172 150	$1 362 100
Benefits							
More Effective Site Investigation	$50 000	3	$16 500	$33 500	$50 000	$50 000	$50 000
Fire Risk Mitigation	$100 000	3	$33 000	$67 000	$100 000	$100 000	$100 000
Flood Risk Mitigation	$9500	3	$3135	$6365	$9500	$9500	$9500
Better Environmental Management	$50 000	3	$16 500	$33 500	$50 000	$50 000	$50 000
Better Lease Management	$100 000	3	$33 000	$67 000	$100 000	$100 000	$100 000
Planning Enquiries and Permit Applications	$43 000	3	$14 190	$28 810	$43 000	$43 000	$43 000
Unidentified Works	$37 500	3	$12 375	$25 125	$37 500	$37 500	$37 500
More Focused adhoc Mapping	$27 300	3	$9009	$18 291	$27 300	$27 300	$27 300
Total Benefits	**$417 300**		**$137 709**	**$279 591**	**$417 300**	**$417 300**	**$417 300**
Cumulative Benefits			$137 709	$417 300	$834 600	$1 251 900	$1 669 200
Net Benefits			−$487 541	$57 591	$252 350	$257 350	$227 350
Cumulative Net Benefits			**−$487 541**	**−$429 950**	**−$177 600**	**$79 750**	**$307 100**

Note: 1 = Salaries plus 30% on-costs (GIS Operator for 6 months in Year 1);
2 = Maintenance estimated at 10% of product purchase;
3 = Benefits taken up at 33% (Year 1), 66% (Year 2) and thereafter at 100%.

Typically the above NPV is usually calculated at the percentage borrowing rate (for that organisation) with another NPV at (say) 2% different to indicate the sensitivity if the percentage rate changes during the life of the project.

In the above case study, the Benefit/Cost Ratio can be calculated as follows:

$$\frac{\text{Cumulative Benefits}}{\text{Cumulative Costs}} = \frac{\$1\,669\,200}{\$1\,362\,100} = 1.225$$

This value of 1.225 indicates a positive Cost/Benefit ratio, i.e. it is greater than 1.

The Payback Period is the point at which the cumulative costs are equal to the cumulative benefits – this can be calculated as 3.6 years from the following chart. The Return on Investment (ROI) is also shown, depicted as the shaded area between the Cumulative Costs and Cumulative Benefits.

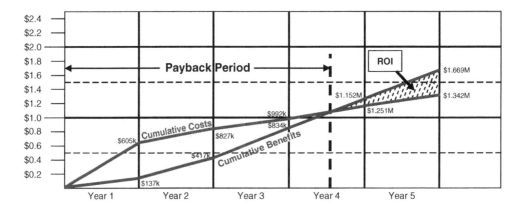

In summary therefore, the above information is usually considered to be all that is required for a 'typical' Cost/Benefit Analysis. If information to this level of detail is presented to the Executive Management or Board of an organisation, then this would generally be enough for them to make a decision to invest (or not) in this technology.

This information shows that the proponents of the Cost/Benefit Analysis, and the attending proposal, have at least 'thought this through' and undertaken a diligent analysis of all the possible financial considerations. Of course, the Executive Management or Board of the organisation may then make a decision based upon non-financial information (perhaps some of the intangible benefits) to either approve or disallow the initiative.

Regardless of the outcome of the Executive Management or Board deliberations, this Cost/Benefit Analysis should become the 'reference document' or 'baseline' for the GIS for future review.

Chapter 1 of this book discussed the notion of 'goalposts for the business', often expressed as Key Performance Indicators (discussed further in Section 11.3). This Cost/Benefit Analysis is such a 'goalpost' – it should describe (at Year 0) the expectations, costs and possible benefits of the GIS initiative. At Year 0+1 it should be possible to undertake a reconciliation of 'actual' versus estimates, similarly for Year 0+2, etc., which should, over time, become a useful indicator of the veracity of the original Cost/Benefit Analysis and the progress being made towards implementing it.

And using this Cost/Benefit Analysis as 'a goalpost', at Year 0+3 (or Year 0+4) it should be possible to return to the Executive Management or Board and prove (in reasonably unarguable terms) that the GIS Strategy was a sound investment and has returned the benefits envisaged.

9.4   BUSINESS CASE

Business Cases are typically extensions to a Cost/Benefit Analysis (such as above) and are often a very specific requirement of the subject organisation. The Business Case is typically a succinct summary of no more than 10 pages and generally provides an overview of the project for the Board or Executive Management Team to base a decision on.

The Business Case structure generally comprises:

- a summary of the existing situation/problem;
- the business opportunity which the Strategy/proposal represents;
- the scope of the work proposed to be undertaken and alternatives considered;
- the costs and benefits (summarised from the Cost/Benefit Analysis);
- the risks and the impacts on the project with commensurate mitigants for each risk;
- the timelines proposed; and
- other issues such as an environmental assessment (if applicable), intellectual property issues, etc.

Most organisations generally use a template for developing Business Cases so that they are in a consistent form and cover all pertinent issues which a Board may want to peruse. As such, each Business Case is unique to the organisation and is typically focused on achieving strategic objectives which are considered important to that organisation.

Typically a Business Case would show the Cost/Benefit section as the summary table and chart (from above), with the full Cost/Benefit Analysis attached as an Appendix, along with the GIS Strategy and any other pertinent documents.

9.5   CONCLUSION

In summary, the Cost/Benefit Analysis process (and subsequent Business Case if required) should be considered as the 'blue-print' for the development of the project.

The Cost/Benefit Analysis usually is undertaken as a component of a GIS Strategy, or immediately following the Strategy, and takes the initiative to a 'finer level of detail' in order to gain Executive Management or Board approval to proceed.

Following approval by the Board, the detailed planning of the project usually commences. This typically involves further (and detailed) investigation and analysis, often leading to modification of the original plan, and consequent modification of the cost estimates contained in the earlier plan.

As such, the Cost/Benefit Analysis (as shown above) continues to iterate and be refined, and at each refinement cycle, depending on the variance of the refined costs from the original, it may have to be re-submitted to the Board for updated approval. If this can be achieved, then the project is well on its way to success. If either the costs have been underestimated (or not estimated) and/or if the benefits are vague and unable to be realised, then the project will surely fail. If this happens there will usually be serious questions asked by management, often with disastrous consequences for those responsible for making this successful.

Tip

It is absolutely essential that the Cost/Benefit Analysis adopt a doctrine of 'full disclosure', not 'glossing over' costs, being 'fiscally very prudent and 'telling it like it is'.

In the analysis:

- the costs must be the absolute total *Maximum* costs that should be incurred; and
- the benefits must be real and able to be realised.

The reason for this is quite simple – if the business case were to be approved and the budget provided (using the costs in the Cost/Benefit Analysis), then there would be an expectation from management that the GIS be implemented in full and the benefits realised.

If the costs have been under-estimated in order to get approval, then this will become apparent when only half a system is implemented and the benefits are not realised. And when that happens, the person who proposed the GIS will look quite silly and may face a severe career move.

---

10 Selecting a GIS

Selecting and purchasing a GIS is usually considered the first major step in the process of an organisation becoming 'GIS enabled'. However, in order to ensure that any subsequent implementation will be successful, this step must be preceded by a proper Needs Analysis and Cost/Benefit Analysis as outlined in previous chapters, otherwise the chance of failure will be high.

Therefore, it is only after all previous steps have been undertaken – the Cost/Benefit Analysis has been approved and a budget has been allocated for the implementation of a GIS and associated works (interfaces/databases, etc.) – that this stage can proceed, i.e. the process of selecting and procuring a GIS system. While this may seem to be the major step in the process, it is just the 'start of the rest of the process' and although it might seem like a reasonably simple task it can be fraught with problems.

It is also at this stage (i.e. when an organisation has decided to purchase a system) that there will usually be a number of sales-persons from vendor companies 'hanging around' the organisation, vying for attention, trying to explain why their product is superior and trying to 'help' the client define their requirements, no doubt to ensure that they will have a much better chance of winning the subsequent evaluation and/or tender.

10.1 INTRODUCTION

Selecting a GIS is no different from any other major purchase – one needs to do some homework, determine what is available, how much it will cost, how will it be implemented (and impact the organisation) and when it can be installed. For a private company, this can be an easy exercise – as simple as placing an order with a supplier. However, for organisations such as government agencies where there is a need to ensure that all purchases are undertaken in a prudent and transparent fashion, arranging for quotations and/or tendering is usually undertaken.

It should be noted that the procurement and tendering practices are generally different in each country. In the EU, for example, there is now a European-wide directive on how tendering is supposed to be handled for different contract values, while in some other countries there is still a lot of room to manoeuvre. Further-more, doing business in developing countries can be highly risky and prospective suppliers would be wise to consult their Foreign Office and/or gain local knowledge before proceeding.

In general though, if the organisation is a government agency or public utility, tendering is generally required for all purchases above a threshold such as $100 000, with purchases less than this value simply requiring three quotes. A tendering process generally requires that it is advertised in the state/territory/national newspapers and that there is a need to

Achieving Business Success with GIS   Bruce Douglas
© 2008 John Wiley & Sons, Ltd   ISBN: 978-0-470-72724-9

have transparency, to abide by probity guidelines, to be fair to all suppliers, to be impartial and to select the most cost-effective solution (noting that this does not always mean the least expensive solution).

Therefore, one of the first questions that the project manager in the organisation should ask is 'Do we need to undertake a tendering process?'. If a tendering process can be avoided, while still selecting the most appropriate GIS system, it can be a more preferable, faster and cost-effective process. However, to not undertake any sort of formal evaluation and selection process can easily lead to the wrong system being selected. This is then the least defensible process, particularly when there could be risks of questions being raised about how a particular system came to be selected.

If vendors detect that the tendering process is not fair and impartial, or that it is not being done correctly, and if they consider that this may adversely impact them (i.e. they might lose the tender), they can (and often do) complain to the appropriate departmental head or politician. When this happens (and depending on the outcome of subsequent discussions and investigations) the Tender Evaluation Team could be reprimanded, re-directed or the tendering process may be re-started with different personnel.

Nevertheless, having said all of this, if the department has to follow a tendering process, but already has a GIS system (perhaps in another part of the department) or there are very strong reasons to purchase a particular GIS system (and these reasons are practical and defensible), then it may be possible to get an exemption from tendering. Usually this is done at the Chief Executive Officer or Director-General level and must be done carefully – the vendors who will lose from this process could complain (as indicated above), with a result that the decision could be overturned.

The corollary to this is that vendors are not silly, nor do they want to waste their own time for no reason. If they realise that the Department has already made the decision to purchase 'Product X', then there is little to be gained from them trying to get that overturned, as a subsequent tendering process will most likely return a decision to purchase 'Product X' after a considerable waste of time and effort by both the Department and the vendor of the other product.

Therefore the decision to 'Tender or not' should not be made lightly. If there is no choice but for the organisation to go through the tendering process, then the only option is to ensure that it is done correctly, that all due processes are followed and that all decisions are documented and defensible.

However, tendering can be a very time-consuming and expensive exercise, therefore it is usually important to 'get it right the first time', i.e.:

- to ensure that the Tender specification accurately reflects that which is required – if the specification does not reflect organisational needs, then the organisation will either have to accept that which is in the Tender specification (when the process is completed) or cancel the tendering process with a view to redeveloping the specification and re-calling Tenders (often with considerable embarrassment);
- to ensure that the budget is adequate to purchase the system as reflected in the Tender specification (while everyone wants to buy a Rolls Royce, few can afford it);
- to ensure that the Tenders are evaluated by a team of people who represent the business needs of the organisation and can correctly and adequately evaluate the tenders without any conflict of interest; but

• most importantly, to ensure that the evaluation team involves people who have done this before – usually an independent consultant specialising in this area will be able to provide the best assistance in this regard while ensuring that all probity guidelines are followed.

Case example

Several years ago, I was the (consultant) systems selection team leader for a $170 million Tender. Because of major concerns over probity and 'due process' (mostly initiated by vendors since one of the competitor vendors already had a system in the Agency), the client department hired a second firm of consultants 'to observe' all tendering processes, particularly the evaluation and selection process.

During meetings it was not unusual to find that one of these guys had quietly slipped into the meeting room and was just sitting in the corner 'taking notes' and making sure that probity was being observed. In essence, while some may consider this as 'over-kill', it was future insurance in case a complaint was made that the process was biased or inappropriate.

10.2   SELECTING A GIS USING A TENDERING PROCESS

Once the decision has been made to go through a tendering process, it must be recognised that there is an implication that this must be done correctly and in accordance with probity rules.

Given that this is accepted, the next decision is whether there is a need to go through an Expression of Interest (EOI) stage prior to proceeding to the Tender stage. EOI processes are usually undertaken as an 'information gathering' exercise in the lead-up to the Tender. That is, if there is a need to further define the technical requirements or to get a better understanding of the capabilities available on the market, then an EOI process can be worthwhile. An EOI process is also useful to provide demonstrations to the staff of the purchaser who may not fully understand all of the capabilities of GIS systems in the market-place. As such, an EOI phase is often referred to as an 'Education Or Information' phase.

EOIs are also useful to 'weed out' those companies that do not have the capabilities (but would like to think they have) for a subsequent Tender evaluation. Therefore EOIs are typically less structured than Tenders. The specification usually contains words such as 'tell us how you do x' rather than 'your system must do x', and because the purpose of an EOI is to select a short-list, not a single supplier, less definition is required in the specification and selection process.

EOIs typically do not ask pricing details, except in broad terms, in the knowledge that any prices supplied will be almost useless, since a contract will not result from the EOI and therefore the vendor will be unwilling to supply prices which may create the perception that their product is too expensive when compared to the opposition products.

Tenders, on the other hand, require more rigour and more evaluation and are therefore more time-consuming. One of the main reasons why an EOI process is undertaken is to derive a short-list of typically no more than three companies from which to request Tenders, otherwise the tendering process can become exceptionally onerous.

However, it should be recognised that unless an EOI is mandatory, it could constitute an over-kill for smaller projects, especially in view of the work involved and of raising vendor expectations. Often the tendering guidelines in the relevant country will state whether EOIs are required for the contract value in question.

THE EOI PROCESS

Once the decision has been made to undertake an EOI process, the following steps should be undertaken.

Develop the Requirements Specification

A specification for an EOI should define (i.e. outline) the business requirements for the GIS, any specific application software (e.g. pipe network modelling) with enough detail to indicate to the vendor the type of functionality required, whether any specific interfaces are required, the provision of a database, etc. At this stage the FRS developed in Section 7.4 should be used as the basis for the EOI.

While some specifications are quite vague, it is always better to provide as much definition as possible, rather than simply to indicate 'the department requires a GIS'. The role of the Business Analyst or Consultant in this process therefore is to further define the business requirements from that which has already been broadly documented in the Strategy and Cost/Benefit Analysis stage, such that these requirements can be used as a specification for the evaluation, e.g.:

- Overview of the Scope of Works, including project deliverables, timelines, conduct and structure.
- Response format, evaluation process, levels of compliance, etc.
- Overall System Requirements – hardware, operating system, database and middleware, end-user environments (if more than one), development environments, software and utilities, user customisation, integration with GPS (if required), data exchange, system integration requirements.
- GIS Requirements – user interface, map viewing/display, tools, query reporting, map production and business output, network tracing, 3D functionality, topology/polygon analysis, raster image integration and viewing, external file association, data input and editing tools, data validation tools, coordinate geometry tools, data modelling tools, data conversion tools, view/query requirements, mobile computing requirements, business intelligence tools.
- Business Requirements – project management, customer service, training, application software development services, data supply applications, etc.

Contractual Specifications

In addition to the requirements specification, a contractual specification is usually appended to the EOI which contains all of the contractual, process and tendering guidelines that may be applicable to that organisation.

Forming the 'GIS Committee'

As part of this process of definition and specification of business requirements, it is important that considerable input be gained from all of the stakeholders in the business groups who will become the end-users of the GIS system, once installed.

In order to get the right level of involvement and 'buy-in', it is common practice to form a GIS Committee consisting of representatives of all stakeholder business groups. This Committee then provides overall direction to the process, ensures that all business groups have sufficient input and reviews/evaluates the responses to the EOI. This also results in this Committee 'owning the process' and therefore the end results (i.e. the decision).

The project manager for the GIS implementation should Chair the GIS Committee and should report to the GIS Steering Committee (see Section 6.3).

Setting Expectations

As part of this continual 'defining and refining' process, considerable care should be taken by the project manager to provide further education to the Committee and to ensure that the expectations of the GIS stakeholders are set at the appropriate level.

One of the problems continually encountered when setting expectations comes from vendors. Because most vendors continually extol the virtues of their software, users' expectations are raised to such an extent that they come to believe all of the 'sales hype'. Therefore one of the tasks of the Chair of the GIS Committee is to ensure that practical and realistic expectations are reinforced, rather than sales hype.

In setting expectations, the maxim of 'under-promise and over-deliver' should always be observed, noting that the converse will always lead to disappointment and frustration with the system. The ideal outcome is that when the system is implemented, the users find that all of their requirements are met and/or exceeded.

Developing the EOI Evaluation Plan

In order to undertake an evaluation of the EOI responses which is fair and in accordance with probity guidelines, an Evaluation Plan should be developed which sets out the process by which this will be undertaken.

An important component of the Evaluation Plan will be developing the scoring method-ology and the weights. It is typical for all components of an evaluation to be scored and that this score be multiplied by a weight for that category (or criteria) to derive the overall 'weighted score' for that category or criteria. The weight is an indication of the overall importance of that category or criteria to the overall evaluation. The following sections discuss typical weighting and scoring methodologies.

It is imperative that the EOI Evaluation Plan (and the weights) be completed and signed-off prior to the EOI being released, and certainly well before the EOI process closes and any responses are received.

All GIS Committee team members should also sign a declaration that they agree with the EOI Evaluation Plan and that they have no conflict of interest prior to the EOI process closing and any responses are received. Of course, if they have a conflict of interest, then they should excuse themselves from the evaluation.

Developing the Weights

A key issue in any evaluation process is developing, and gaining agreement on, the weights that will be used for the evaluation. Of course, all weights should total 100%, but which component of the evaluation is the most important? For example, how important is it to have a supplier with a local support capability? How important is it for the supplier to fully meet all of your software requirements? All else being equal, will you tolerate a few shortcomings in one area of the overall solution?

Therefore developing these weights, or 'importance criteria', is essential. And as part of this process of developing the weights, care should be taken not to apply excessive weight to technical functionality. The long-term viability of the vendor as a supporting organisation, for example, can be an important aspect, so the commercial considerations should deserve some importance.

The tables below show some examples of weights used for past evaluations and how the criteria for the weights are segmented. In the following table, the Technical and Professional Competence of an EOI response is given 75% of the overall score and Commercial Considerations are allocated 25%.

Area of Evaluation	% of Total Evaluation
Technical & Professional Competence	75%
Commercial Considerations	25%
Total	**100%**

This can be further dissected in the following table, showing that compliance to the technical specification is allocated 45% of the total overall score (after weights have been applied), while 'Compliance to Conditions of Contract' (for example) rates 5%.

Sub-area	Description	% of Total Evaluation	Totals
Technical & Professional Competence			
1.1	Technical (compliance to specification)	45%	
1.2	Organisational Considerations	30%	
Sub-total			**75%**
Commercial Considerations			
2.1	Financial Viability	5%	
2.2	Risk & Insurance	5%	
2.3	Compliance to Conditions of Contract	5%	
2.4	Conflict of Interest	5%	
2.5	Employment Standards	5%	
Sub-total			**25%**
Total			**100%**

Undertaking a further refining of Technical Competence (1.1 in the above table) shows, in the following table, that 'Performance Characteristics' (for example) are given an overall weight of 4.05%. Of course, all weights must total 100% at all times.

Therefore, the development of the weights 'sets the scene' for 'how important' differing criteria are against each other. And because everyone has a different understanding of 'what

Sub-area	Description	% of Total Evaluation	
1.1	*Technical (compliance to spec)*		
1.1.1	Functional Characteristics	8.55%	
1.1.2	Cost Effective Deployment & Flexible Interfaces	8.55%	
1.1.3	Performance Characteristics	4.05%	
1.1.4	Technical Characteristics	8.55%	
1.1.5	Requirements for Services	8.55%	
1.1.6	Project Management & Development Methodologies	2.25%	
1.1.7	Training	2.25%	
1.1.8	Ongoing Vendor Support	2.25%	
Sub-total			**45%**
1.2	*Organisational Considerations*		
1.2.1	Capability	5.4%	
1.2.2	Past Performance & Current Work	12.6%	
1.2.3	Quality System for Deliverables	2.4%	
1.2.4	Customer Service	2.4%	
1.2.5	Strategic	2.4%	
1.2.6	Innovation	2.4%	
1.2.7	Organisational Structure	2.4%	
Sub-total			**30%**

is more important', this is often the source of considerable debate on the EOI Evaluation Team.

It should be noted that the Evaluation Plan and the weights are to be used as an internal document only. That is, while the EOI might discuss the evaluation criteria, the weights to be used in the evaluation should not be released to the vendors.

Developing a Scoring Regime

The EOI Evaluation Team typically would also adopt a numerical scoring scale such as the one in the table below, against which each team member will assess all the evaluation criteria to which they have been assigned.

Rating	Description	Score
Excellent	Exceeds requirements in all ways, with very little or no risk	9–10
Very Good	Meets requirements in all ways and exceeds it in some, with little risk involved	7–8
Good	Meets the requirements, is workable and is an acceptable risk	5–6
Marginal	Nearly meets requirements and is workable but may be deficient or limited in some areas, with some element of risk	3–4
Poor	Offer is sub-standard, is difficult to assess against criteria or is a high risk	1–2
Non-compliant	Offer has stated non-compliance, demonstrated non-compliance or there is insufficient information to assess	0

During the evaluation this scoring regime will be used as a guide (for scoring) and then will be multiplied by the weight of that category to calculate a weighted score for each EOI category.

Releasing the EOI to the Market-place

The EOI document should be advertised and released to the market-place for a period of (typically) 4–6 weeks with a requirement that the responses be provided in a hard/soft copy format by a specified date.

This is the period when the vendors put together their response. During this period, all enquiries by vendors should be directed to the Chair of the GIS Committee.

EOI Briefing Session

In some instances, an EOI Briefing Session may be useful to present and explain the requirements of the EOI to 'the industry' in a concise manner, as well as to field questions from vendors. While the advantage of this process is that all vendors get 'the same story' at the same time, the disadvantages often outweigh this, e.g.:

- vendors are often reluctant to ask searching questions in the presence of their competition, preferring to do this in a one-to-one meeting with the GIS Committee, thereby still requiring additional meetings, and defeating one of the main purposes of the Briefing Session; and
- the Briefing Session requires some amount of effort for both the tendering organisation and the vendors, often only to re-state that which is in the EOI document.

Receipt and Evaluation of EOI Responses

By the nominated date, the responses to the EOIs should be received. It is common practice for a typical EOI process for a GIS system to attract between 10 and 15 responses from vendors, depending on the complexity of requirements and applications requiring to be supported. Therefore, following the close of the EOI process and receipt of the responses, the EOI Evaluation Team must evaluate these responses in accordance with the EOI Evaluation Plan. The usual steps in this process would be to:

- read and score all responses based upon the submitted documentation to determine a list of vendors from which demonstrations would be requested;
- attend demonstrations of vendor products to verify capability as stated in the EOI response;
- update the scores of each response based on demonstrations;
- determine an overall 'weighted score' for each system; and
- develop a short-list of vendors for the Tender process.

Prepare an EOI Report

One of the final steps in the EOI process is generally to prepare an 'EOI Report' for the GIS Steering Committee containing a commentary on the process used and recommendation(s) on the short-list of vendors from which to invite Tender responses.

Following acceptance of this report by the GIS Steering Committee, it is normal practice to advise all vendors of the selected short-list and that the unsuccessful vendors be invited to attend a de-brief session if required.

De-brief Sessions

Most unsuccessful vendors usually want to know why they lost – was their price too high, did they not have the functionality required, was their service too poor, etc. Therefore offering a de-brief session to each of the unsuccessful vendors is often a good way to provide this information and to 'close out' the process.

But *beware*, care must be taken that this session is not used by the unsuccessful vendor to complain about the winning vendor. It should be made clear at the outset by the GIS Steering Committee that they will only discuss the reason why the unsuccessful vendor lost, not why the winning vendor won.

In my experience, vendors are often very appreciative of this feedback on their performance during the EOI process and it provides useful information for them to ensure that they do better next time. It may also be useful to put some of the detail in writing, particularly sensitive material, so that there can be no future argument as to what may have been said.

A de-brief session is also a nice way of saying thanks for the vendor's time and effort, particularly in jurisdictions where they may not be able to claim any expenses for the Tender.

Conclusion of the EOI Process

This concludes the EOI process. If undertaken successfully, the outcome should be:

- a short-list of suitable vendors selected in accordance with the specification of requirements with which to proceed to Tender;
- an understanding of the GIS capabilities in the marketplace by the GIS Committee; and
- a better understanding by the GIS Committee of their requirements for a GIS, and how these requirements could be satisfied by each of the short-listed vendors.

THE TENDERING PROCESS

The Request for Tender (RFT) process is very similar to the EOI process only 'more so', to a greater level of detail and with more stringent contractual, probity and legal considerations. While this section describes some of the processes and issues to be considered in undertaking a Tender, it would be advisable that local regulations are checked before proceeding to Tender.

If the above EOI process was not used and the RFT process is the 'starting point', then most of the steps above should be followed but in the context of an RFT rather than an EOI, and with the enhancements outlined in the steps below.

However, the previous EOI process, being properly run and managed, should have resulted in:

- an understanding of the GIS capabilities in the market and how the organisation's requirements could be satisfied by the short-listed vendors;
- a better definition of the organisation's GIS requirements;
- a smoother Tender process in a more constrained manner.

However, before setting off on a Tender process, care should be taken to ensure that the tendering guidelines being used are in keeping with those required for the particular country, in that in different jurisdictions vendors have different rights. For example, in some countries (such as England) vendors have a right to bill the client for their expenses during a major Tender whereas in other countries (such as Scotland, Holland or Australia) this is not the case. However, I have always found that it is useful for clients and vendors to agree at the initial stage whether vendors' expenses are to be covered, particularly if one is asking them to undertake a substantial amount of development to be able to demonstrate their capability.

The following steps are normally undertaken in order to ensure that the Tender process meets probity, contractual and 'value for money' requirements.

Developing the Tender (by Updating the EOI Requirements)

The GIS requirements specification, as documented for the EOI process, should be updated for:

- the GIS system capabilities as observed during the EOI demonstrations; and
- the GIS requirements (of the organisation) as further refined from observations of capability during the EOI process.

As such, the Tender specification should be a 'tightening up' of the previous EOI specification and should represent the full and final requirements for a GIS, which if implemented would (should) meet the business needs of the organisation.

All requirements should be specified as Mandatory, Desirable (with the Desirables further discriminated into Highly Desirable, Desirable, etc.) or Comments, to indicate compliance requirements to the vendor. Note that the following would be typical:

- Mandatory: no weights are assigned to mandatory requirements – they either comply or do not comply (i.e. a Yes or No response) – but note that if a tendered solution does not meet a Mandatory item they have failed (if that item really was mandatory)
- Highly Desirable: e.g. weight of 6–10.
- Desirable: e.g. weight of 2–5.
- Comments: not scored or weighted.

Note that extreme care should be taken with allocating 'Mandatory' to items, remembering that this will be published in the Tender specification and will be scrutinised closely by all vendors. Therefore, if a Mandatory really is Mandatory, the vendor who does not meet this requirement is automatically excluded and has failed the tendering process.

At the risk of re-stating this point – a Mandatory really must be Mandatory. The Tender Evaluation Team must have very good (and defensible) reasons why a particular functional requirement really is mandatory, why it is absolutely essential to be mandatory and why a vendor who does not meet this requirement must be excluded.

> **Warning**
>
> If, during a Tender evaluation, it is decided that 'for this vendor, mandatory is not really all that mandatory', there will be extreme angst from the other vendors who discover that their competition has been given an unfair advantage. A complaint by the losing vendor will almost certainly ensue and will result in substantial explanation being required to be made to the Chief Executive Officer or Departmental Head as to how this breach of probity was allowed to occur.
>
> Therefore, a Mandatory requirement really must be mandatory. That is, if a vendor fails a mandatory they are excluded from further consideration and have lost the Tender.

Some examples of issues which could be mandatory (note that all mandatories generally use the word 'must' rather than 'should' or 'could') are:

- The tendered system must run on Windows XP (*i.e. a solution based on Windows 2000 or Linux is not acceptable and that Tender will fail and be excluded from further consideration*)
- The tendered solution must store all geospatial data in the company's RDBMS (*i.e. a solution using a different RDBMS is not acceptable and the Tender will fail and be excluded from further consideration, similarly for Tenders which store geospatial data as graphic files*)
- The vendor must provide support in the company's city (*i.e. support based in another nearby city is not acceptable and the Tender will fail and be excluded from further consideration*).

Again, to re-state that which should be becoming obvious – a Mandatory really must be mandatory.

Developing the Pricing Schedules

A key focus of any Tender is, of course, the price. As obvious as it seems, it is important to ensure that an 'apples and apples' comparison is made between the prices submitted from different vendors, so that an overall understanding can be made of the 'total cost of ownership' of one tendered solution compared to the next.

Some vendors have a practice of offering low purchase prices for their software but then charging very high prices for maintenance and other services, particularly software development. This is so that their product appears cheap to buy but a 'total cost of ownership' calculation over a 3- or 5-year period highlights that their solution is much the same price as their competitors, or in some cases may be more expensive.

In order to ensure that all vendors answer all pricing issues and that 'grey areas' are not left open, it is normal that pricing schedules be developed and included in Tender specifications for completion by the vendors.

The development of these pricing schedules (as spreadsheets) should be based on the assumption that 'if a price is not asked for, the vendor will not provide it' given that all vendors (in my experience anyhow) do all that they can to minimise their price so as to win the business, on the assumption that they can increase their prices later to include all the 'hidden extras' that were not outlined before.

To avoid this situation, it is useful for all price schedules to carry a banner notice with words to the effect that if the price is not included in the schedule, then it will not be included in the contract and the vendor will not be paid for that component, but the vendor will still be expected to provide a working system as tendered.

A Pricing Schedule is a spreadsheet listing all of the possible components that will (or may) be supplied against which the vendor should put a purchase and annual maintenance price.

Developing the Tender Evaluation Plan

The previous EOI Evaluation Plan can now be upgraded to become the Tender Evaluation Plan, reinforcing the process of how the evaluation should be undertaken and including a review/update of the scoring and weighting process to ensure that it is suitable for the tendering process.

As in the EOI process, it would be expected that the Tender Evaluation Plan (and the weights) be completed and signed-off prior to the close of the tendering process and before any responses are received. The Tender Evaluation Plan should also include an outline of how Bench Tests are to be conducted (if required) and how Reference Checking and Site Visits will be undertaken (refer below).

It is customary that all Tender Evaluation Team members have some training in the duties that are expected of them and that they sign a declaration that they agree with the Tender Evaluation Plan, that they have no conflict of interest and they agree with the probity guidelines. This must occur prior to the close of the tendering process and before any responses are received.

Case example

Several years ago, I was Chair of a Tender for a GIS project for a large multi-function government agency. One of the members of the Tender Evaluation Team broke probity and discussed confidential Tender information with her supervisor (who was not involved with the Tender). He then tried (unsuccessfully) to change the outcome of the team decision and have one of the tenderers dis-allowed. The Probity Officer had attended the Tender Evaluation Team meetings and was aghast that this was happening. Clearly this was inappropriate and bordering on illegal.

The team member was remonstrated by the Probity Officer and agreed that she did break probity. Because there was a proper probity process being used, with all meetings attended by the Probity Officer (and decisions minuted), the Contracts Manager ruled that the tendering process did not have to be re-started (and could continue) as long as the team member did not discuss confidential information outside the team again. She agreed.

While there were obvious rumours about the reason why her supervisor may have wanted to have a particular tenderer excluded so that another tenderer would win, this was not investigated by the Department, much against my recommendations.

Contractual Specifications

The contractual specifications to be included in the Tender should be quite specific, and should again be reflected in a booklet appended to the technical specifications. The contract specifications must contain all of the contractual and tendering guidelines that may be applicable to that organisation in that jurisdiction and should include a blank copy of the contract that the vendor will be asked to sign, should they be successful with the Tender.

Re-focusing the GIS Committee as the Tender Evaluation Team

The progression from the EOI to the Tender will provide the basis for the GIS Committee to be re-focused (re-badged) as the Tender Evaluation Team, usually with similar (or the same) business unit representation.

Again, the project manager for the GIS implementation should Chair the Tender Evaluation Team and should report to the GIS Steering Committee.

Re-focusing Expectations

Following the EOI process, and as part of this continual 'defining and refining' process to upgrade the EOI specification to that of a Tender Specification, the project manager should continue to ensure that the 'expectations' are set at the appropriate level in preparation for the Tender selection.

Releasing the Tender to the Short-list

The Tender document should be released to the short-listed vendors (from the EOI process) for a period of approximately 6 weeks, with a requirement that the responses be provided in a hard/soft copy format by a particular date. During this period, all enquiries by vendors should be directed to the Chair of the Tender Evaluation Team and all enquiries should be in writing. It is a useful practice for all Tender Evaluation Team members, particularly the Chair, to maintain a diary of all conversations /communications with anyone external to the Tender Evaluation Team, particularly vendors, about anything relevant to the tendering process.

All Tender Evaluation Team meetings should be minuted and in some cases attended by a Probity Officer, usually a contracts officer.

Receipt and Evaluation of Tender Responses

While the Tender evaluation process would initially follow the EOI process quite closely, it usually consists of a number of segments, all of which must be documented in the Tender Evaluation Plan.

The first stage of evaluating Tenders is to undertake a Preliminary Evaluation to understand the tendered solutions and to review the proposed solutions in order to determine which tenderers will be invited to undertake Bench Tests.

The following methodology is often used for the Preliminary Evaluation:

1. Evaluation Team members read all submitted Tenders and score all evaluation criteria to which they have been assigned. Such scores are based upon the tendered 'written

document' only and not influenced by the 'known capability' of each solution. The Tender Evaluation Team members would use a score sheet for each Tender Response document – this is undertaken by each team member in isolation from other team members.

2. Once these are complete, all evaluations are forwarded to the Chair for compilation into a preliminary composite evaluation score sheet (one per Tender Response).
3. The Evaluation Team then reviews these composite assessments and determines:

   a. which solution(s) clearly cannot meet the organisation's requirements; and
   b. which solution(s) could meet the organisation's requirements, subject to testing the veracity of that solution's capability.

4. Those tendered solutions which fall into the second category are then invited to undertake a Bench Test to test the veracity of their proposed solution.

Bench Testing the Tendered Solution

The purpose of undertaking Bench Tests is to ensure the veracity of each tendered solution, in much the same manner that one would take a car for a test drive before making the final purchase. During the Bench Tests, each team member would normally update their preliminary scores by observation of the capability demonstrated during the Bench Tests.

In order to ensure that each Bench Tests is fair and that the test can provide the means of evaluating the differences between each system in a working environment, it is recommended that the following process be undertaken:

1. The aim of the Bench Test is to test the proposed solution in an environment which would simulate the actual working environment should that solution be selected. Therefore, it is usual that the Bench Test would consist of a subset of 3–5 typical business processes that the system would be used for should it be selected and implemented.
2. Each of the 3–5 typical business processes should be documented (scripted) such that the steps/tasks required to be undertaken are described. This is usually pre-defined and provided to each tenderer so that they can practice and are aware of the tasks required to be undertaken and therefore have no excuse that they were not prepared to demonstrate a particular capability.

Case example

A large tendering process that I was managing (which followed this process) resulted in one tenderer not bothering to practice with the data provided. During the bench test, a fatal flaw was shown in his software when he could not find a street address because the data provided was not 'structured' as his software required. All tests went on hold while he ran down to his rental car to retrieve his street directory so that he could proceed. This demonstrated that if this system was selected, it would be necessary to 'hardwire' the data to the software, at considerable additional expense, thereby making data updates very difficult.

3. Data (spatial and attribute) should be provided to the tenderer so that the Bench Tests are undertaken using actual working spatial data from the organisation. This is particularly critical in a GIS system evaluation, given the proclivity of some GIS solutions to 'hard wire' spatial data to their solution and to optimise that data for performance prior to a demonstration.

4. Each tenderer should be allocated a time to undertake the Bench Test; typically such a test may take half a day for a small Tender and up to 1–2 days for a Tender with a large set of requirements.

5. In order to facilitate this process, a 'Bench Test Package' should be created and provided to each tenderer prior to the tests being undertaken – such a package would include:

   a. an outline of the tasks that are required to be undertaken – outlined as the 3–5 typical business processes;
   b. the spatial and aspatial data necessary to undertake the required tasks and to simulate loadings on the system;
   c. any other instructions considered necessary for the successful undertaking of the Bench Tests; and
   d. an agenda for the tenderer to work to in order to ensure that the tests are completed.

Tip

A favoured trick by vendors is to 'run out of time' if they cannot do some of the tasks that they have tendered that their software can do. That is, they may have embellished (lied about) the capabilities of their software in their Tender response, and now that they have to demonstrate the capability they try different tricks (such as running out of time) to avoid doing so.

Therefore it is always useful to highlight that, in fairness to all tenderers, if a tenderer fails to complete a segment of a Bench Test then they will be scored zero for that particular segment. This generally works to 'get their attention' and keep them 'on track' to ensure that they do engage the process seriously.

---

6. The Bench Test Package should be provided to each tenderer so that each has the same notice between receipt of the package and the date when their tests are planned to occur.

7. During the Bench Tests, the Evaluation Team members should update their preliminary scores for the evaluation criteria to which they have been assigned.

8. The conduct and outcome of the Bench Tests is usually a major component of the final Detailed Evaluation Process.

Because the Bench Tests will form a major component of the technical evaluation, it is important that the evaluation team observes the conduct of the tests in a completely impartial manner.

Considerable care should be undertaken to avoid trying to help vendors when it becomes apparent that they may not be undertaking the set task in the most efficient manner possible. The role of the evaluation team is to observe and to evaluate, not to help the vendor complete the test.

Separate Costing Evaluation

An option that is often employed in large procurements, particularly where there are a lot of differing business requirements, is to undertake the evaluation of the tendered price as a separate process to the evaluation of the technical issues. Indeed, in some jurisdictions legislation it is mandatory to separate price and technical evaluations.

When there is a separate costing evaluation, it is usually undertaken by using two teams (and separate processes occurring in parallel):

- a Technical Evaluation Team – with a role to evaluate the *technical* issues only; and
- a Price Evaluation Team – with a role to evaluate the *price* only.

The Technical Evaluation Team typically comprises technical business unit representatives, whereas the Price Evaluation Team typically comprises finance representatives.

In order to facilitate this process, the Tender would typically request that the Tendered Price Schedules be submitted in a separate sealed envelope from each vendor so that the tendered technical information is evaluated by the Technical Evaluation Team and the tendered pricing information is evaluated by the Price Evaluation Team. A consequence of this process is that the Technical Evaluation Team is not privy to the costs tendered.

A key benefit of separating the price and technical evaluations is that the technical evaluation team members are not influenced by a solution being perceived as 'too expensive' or 'too cheap'. That is, each technical solution is evaluated uncoloured by pricing considerations and, as such, is evaluated on technical merit only.

Note, however, that it is also useful for the same person to be Chair of both teams, thus facilitating sufficient 'cross-over' to ensure that the cost evaluation, proceeding separately, is undertaking a similar comparison, i.e. that the cost models are based on an 'apples and apples' comparison across the different bids.

The Price Evaluation Team would typically ensure that their cost models resulted in 'whole of life' costs for each vendor, usually over a 3–5-year period, this balancing out the impact of purchase price compared to maintenance price, staffing costs, training costs, etc.

Reference Checking and Site Visits

Once the evaluation process has reached a stage where a probable winning vendor is emerging, it is prudent to undertake some checking of references and to visit other sites where the same or a similar solution has been implemented.

Case example

A tender selection process that I was undertaking several years ago involved doing reference checking on the preferred vendor – the vendors had provided contact details of their references. When I called one of the vendor's referees, his response was 'never heard of them'. Further checking showed that this was indeed true. The vendor thought we would not check references and gave totally fictious referees. Needless to say, that vendor did not win the business.

As part of this process, it is important that all reference checking be coordinated by the Chair of the Tender Evaluation Team and that a standard list of questions be prepared and asked of all referees to be checked. It is also prudent that all team members be present for the reference check.

It is normal for initial references to be undertaken by (speaker) phone and all discussions to be documented. Following this, a small number of reference sites may be selected to visit, again coordinated by the Chair of the Tender Evaluation Team. Note that the vendor should *not* be present during the site visit in order that a full and frank discussion can be undertaken at the reference site about the capability and performance of the software, vendor, maintenance, etc.

Site visits should also be arranged with sites which have not been provided as reference sites by the vendor if the Tender Evaluation Team are concerned that there are issues which need to be addressed at those sites.

Conclusion of the Tender Evaluation

The final step in the Tender evaluation is for the Tender Evaluation Team to bring together 'all of the pieces' so that an informed decision can be made about the following for each of the proposed solutions:

- the weighted score of the technical capability;
- the 'whole of life' cost;
- the professional capability of the vendor;
- the risk (and the mitigants of those risks);
- the capacity of the vendor to support the proposed implementation;
- the implementation timeline/plan; and
- any other considerations that may be considered relevant.

When the evaluation is complete, a list of the 'first', 'second', 'third' can usually be derived for:

- the best 'Technical' solution;
- the most 'Cost-effective' solution;
- the least risky solution.

This process is very effective in providing reasoned argument as to which is the most preferred vendor (and least preferred vendor) for subsequent discussion and deliberation.

Preparing a Tender Evaluation Report

One of the last steps in the Tender evaluation process is usually to prepare a 'Tender Evaluation Report' for the GIS Steering Committee, the outcome being a recommendation of the vendor to be selected with which to commence negotiations. The Tender Evaluation Report should include:

- background to the process;
- the evaluation process;
- the ranking of Tenders;
- proposed methods for management of perceived risks and issues;
- recommendations for endorsement (i.e. Tender selected for negotiation process).

Any member of the Tender Evaluation Team can submit a minority report, on any aspect of the process, for consideration by the Chair and the Steering Committee if so required, with recommendations for addressing issues raised within such a report.

Again, following acceptance of this report by the GIS Steering Committee, it is normal practice to advise all vendors of the selected vendor (i.e. the winner of the Tender) and that the unsuccessful vendors be invited to attend a de-brief session if required.

THE BAFO PROCESS (OPTIONAL)

An optional process employed on some large Tenders is to request a Best and Final Offer (BAFO) from the final tenderers as a closing stage. This is usually undertaken just before the Tender is awarded and when the short-list has been reduced to two vendors and the selected tenderer is still not finalised. This process is used quite simply as a mechanism to obtain a better price.

While some would argue that this is an unnecessary step, it can serve to reduce the price. As such, it is a common practice, and in some cases is expected by vendors. Therefore some vendors have a practice to 'hold back' their best price until the bitter end, knowing that there will be a final offer required and they 'need to keep something in reserve to offer'. This is their last chance to win the business after the detailed technical process has been undertaken and they need to have a 'carrot to dangle'.

Others would also argue that this is not appropriate in that 'the tendered price is the tendered price' and should not be subject to later discounting. However, it is a practice often employed in large government tenders and is generally taken as being within probity guidelines. Of course, when private companies purchase a GIS, they haggle all the time.

10.3   THE FINAL STAGE OF THE SELECTION PROCESS

At the end of this often long and painful process, whether it be undertaken by direct quotation, EOI + Tender or just the tendering process, there is always a need to enter into a contractual arrangement with the vendor for the goods and services offered.

As such, and following the acceptance of the recommendation to award the Tender/ project to the selected vendor, it is usual that a process commences to negotiate and finalise a contract which can be signed-off by both parties. This is usually a time when the selected vendor would update the implementation timeline, finalise the personnel to be employed on the project, finalise the training programmes and finalise the data conversion (if required). In addition, the purchaser may finalise some of the options proposed so that the final costs can be determined and the contract can be finalised and signed.

A key issue to be considered in the contract negotiations process is the need to contain variations, mostly caused by a combination of the following two processes:

- vendors may have 'low-balled' the price to win the business with the intent to extend (or vary) the project once they have been selected – this results in them gaining agreement for variations (often at a higher scheduled rate than tendered) which are then used to raise the price to that which would give them a higher level of profit (and may have precluded them from being selected if these costs had been disclosed earlier); and

- because a large number of GIS Tender specification are quite vague, a very large number of GIS and IT projects usually need to have the scope varied by the end of the tender process in order to have the required capabilities implemented – again providing 'fertile feeding ground' for vendors to extract variations.

Case example

After contract negotiations and award on a large project, the software developer was required to create an integrated Map Query Interface (MQI), using data from several other application sources using multiple coordinate systems.

The developer completed the MQI which produced data in the coordinate reference system to be used in conjunction with data from other jurisdictions, but the MQI did not supply the datum shift (approx 200 m) to the jurisdictional data so that the data from the two different sources could be shown in the MQI and viewed together correctly geo-referenced.

One would have assumed that the GDA projection would (by default) come with the datum shift so that it could be used with the other projection data in the MQI, otherwise it would be useless in comparing data sourced from the two jurisdictions with different coordinate reference systems.

The developer argued that this was an 'optional extra' and used it as the basis for requesting a substantial variation by asserting that he only needed to supply data on the final projection, even though he knew the solution was unworkable in the manner that he supplied it. Unfortunately, and much against my advice, the departmental manager preferred not to argue the point and paid the variation.

Therefore good project management should ensure that:

- the scope (as expressed by the specification of requirements) is 'quite tight' so that the need for variations by the vendor is limited or excluded; and
- the pricing schedules in the tendering process are 'sufficiently rigorous' so that there are no ambiguities which may give vendors the need to vary their tender prices.

A final process is to develop an 'Implementation Team', usually a combination of the vendor and client representatives who will work together to implement the system, develop additional software (if required), undertake the training, do the data conversion (if required), etc. Therefore, with the 'Selection Phase' ending, the 'Implementing Phase' is ready to begin.

11 Implementing GIS

When the GIS selection process has been completed and a successful GIS solution has been selected, the real work will commence. At this stage, it is important that the successful vendor and the purchasing organisation consider the selection of the GIS as just the start of (what should be) a long and successful working partnership – hopefully very successful for the purchasing organisation as well as being profitable for the vendor.

In commencing the implementation it is important to 'start like you mean to go' and to develop a very structured and methodical way of going about the implementation and 'setting to work' of the system. The steps usually involved in this stage are as follows:

- staff training;
- data capture and/or conversion;
- defining the KPIs (goalposts) for successful implementation; and
- implementing and setting to work of the GIS.

In addition, it may be prudent to consider several long-term stages, such as:

- undertaking a Post Implementation Review; and
- benchmarking.

11.1   STAFF TRAINING

Section 6.5 discussed some GIS training issues. Thorough and competent training is essential to the success of GIS and no amount of training is 'too much'. Training and 'follow-on mentoring' should be undertaken in a methodical and structured manner with a careful record kept of the courses undertaken by each member of staff and the skill level obtained, to be correlated later against work-place observation of competence. In this manner, staff skill levels should be maintained at a 'peak condition' so that all tasks can be undertaken in an efficient manner using the optimum capabilities of the GIS system.

> **Case example**
>
> Several years ago, I was asked to review and manage an existing GIS environment which was having substantial difficulty in meeting deadlines. After a period of observation, I sat with several operators and asked them to 'talk me though' what they were doing. It did not take long to realise that while they were trying to do the best they could, they had very little understanding of what they were doing. I found out later they had been hired from the unemployment queue and had almost

> no training or skill background in the industry, the previous manager preferring to spend all of the budget on software and 'save money' on training.
>
> Needless to say, the money 'saved' from not hiring appropriately skilled staff and not giving those staff adequate training was wasted many fold over through lost time and project over-runs.

For the last 6 years, the Australian and New Zealand GIS industry have consistently only spent 4% of an organisation's total budget on training (see Section 2.5). Most would consider that this is far too low to maintain staff performance at optimum levels.

Training should be context-sensitive – that is, it should be relevant to the work processes of the organisation (preferably using civil data for training civil staff, using council planning data for training council planners, etc.), so that the trainees understand the application of the software to the work environment to which they will employ the software.

Training should also be at the 'appropriate level' of the staff – it is pointless giving advanced training to staff who need an 'Introduction to GIS' course, or vice versa. Similarly conversion training may be required if the GIS is replacing a different GIS and staff will need to be re-skilled from one GIS to another.

That is, a Training Plan should be developed outlining that there could be up to a dozen different courses required to cover all of the different disciplines and to cover all appropriate training levels required (in each of those discipline areas). In addition to GIS training provided by the vendor, there are now companies which specialise in providing detailed and context-sensitive specialised training for different GIS environments.

In summary, any money spent on training is 'money well spent' and if any part of the budget needs to be trimmed, my advice would be to *not* to reduce the amount spent on training but in fact increase it!

Fact

No training = False economy
Would you even think of not training aeroplane pilots or even allowing people to drive without a driving licence? So then why do managers think that their staff can learn GIS by themselves? The implementation of a $1 million GIS without providing adequate training has just wasted $1 million.

Note that while planning for the training can commence before the finalisation of the GIS selection, the types of training may be different for different GISs and usually are not finalised until the GIS is selected.

As indicated earlier, training must be relevant and staff must be able to apply their training as soon as they are back in the office, otherwise it will be forgotten when they need it. One of the most common mistakes is not to differentiate between the different training needs of back-office (database) and front-office (map production, spatial analysis) staff. There is no 'one-size-fits-all' in training.

11.2   DATA CAPTURE AND/OR CONVERSION

Chapter 5 discussed the significance of corporate spatial data, highlighting that GIS is data-centric and that without good quality data the GIS is 'next to useless'. Therefore, sufficient planning should be given to ensuring that adequate data is provided for loading into the GIS so that it can be operational as soon as possible.

Section 5.3 discussed data architectures and the need to plan for the deployment of the data as well as the GIS software for use across the organisation. That is, different deployment strategies may (or may not) impact on how the data is structured and/or stored. Therefore, before any data capture and/or conversion is undertaken, it is sensible to address the data architecture.

Section 7.3 discussed some data issues in the context of developing the GIS Strategy. Almost all organisations have a plethora of data available which can be used in the GIS, but the main focus should be on deciding which data is useful and therefore should be brought forward into the GIS, and which data will probably not be useful and should be left behind (and either archived or only brought forward as and when required).

Needless to say, a Data Capture/Conversion Plan should be prepared before any action is undertaken. This plan should have the input and endorsement of all stakeholders within the organisation as well as those who might be external providers/users of data.

It is normal that this process be initiated by forming a 'Data Consultative Committee' to determine the:

- data classes required (e.g. cadastre, trees, roads, kerbs, pavement, building footprints, valves, pipes, poles, etc.);
- attributes required for each data class (e.g. tree height, genus, age, trunk diameter, maintenance, etc.);
- spatial precision required for each data class (i.e. sub-metre, 1 cm, nearest 10 m);
- data capture method for each data class (e.g. GPS, digitise from plans, aerial photography, field inspection etc.);
- data currency for each data class (e.g. data is updated every 6 months, every visit to each asset, etc.);
- data use for each data class (e.g. for planning purposes, for construction, for reporting, etc.).

Some of this data (e.g. cadastre) will be able to be purchased under licence from a relevant government department in that jurisdiction with provisions for regular updates of that data (e.g. quarterly). However, many jurisdictions have different licensing terms which may prohibit data purchase, and the data may need to be re-captured. Other data may need to be converted from internal plans or maps, while some data may need to be captured in the field.

For each data class, a specification should be developed for the work which will be required to be undertaken to capture and/or convert that data class to that which can be used in the GIS. These specifications can then be used to either manage the process (if it is undertaken using internal staff) or to form the basis of selecting a company to undertake it on contract (using a tendering process).

There are a number of private companies in the market-place with extensive experience in capturing and converting data who can be contracted to do this work. Note that the selection of a company to do this work should be undertaken using a tendering process similar to that outlined in Chapter 10.

However, when such a specification is developed, it should outline in some detail the format of how the data is to be provided, the structure of the data (e.g. layering, line-style, symbology, attribution, etc.), the connectivity of the data (e.g. that all 250-mm pipes be connected and 'form a topologically correct pipe network' which is recognised as such in the target GIS) and how the data will be provided to the organisation.

Note that it is prudent that a quality assurance (QA) procedure be initiated to 'receipt' the data provided by the contractor and to ensure that it has been captured in accordance with the specifications. It is also useful not to process any invoices until after each data package has been 'accepted' as having passed the QA. This may also involve withholding a component (e.g. 10%) of the total contract price until after all data has been accepted and proved to be to the level of quality and completeness required under the contract specification.

It should be stressed that the cost/benefit analysis developed for the Strategy should leave enough margin of uncertainty for the data capture costs, as they will not be confirmed until after the contract negotiations.

Note that while the planning for the data capture and conversion can commence before the finalisation of the GIS selection, some GISs require that data be structured quite differently. Therefore, the finalisation of the data capture and conversion plan, including any Tenders relating thereto, is usually not completed (or commenced) until after the GIS is selected.

Case example

A few years ago I was developing a Data Capture Tender specification for a major power utility while the evaluation of the GIS Tender (which I also developed and was managing) was proceeding. The three systems short-listed for the GIS Tender each required a different approach to the Data Capture specification, which would be reflected in the subsequent Data Capture Tender price.

The timing was such that the client could not wait until after the GIS Tender selection was completed to release the Data Capture Tender, so we had to devise a process where we could do an 'apples and apples' comparison for the Data Capture Tender while the outcome of the GIS Tender was still not finalised.

The solution was to release the Data Capture Tender containing the short-listed solutions from the GIS Tender and advise each tenderer that they needed to price the data capture up to the point where they would have the data ready to load into the GIS (when known). This allowed them to review all the data processes and proceed through the tendering process to the point where they would finally price it. Just before the Data Capture Tender closed, we advised them of the outcome of the GIS Tender so that they could price the Data Capture Tender for loading into that GIS.

11.3 DEFINING THE KPIs (GOALPOSTS) FOR SUCCESSFUL IMPLEMENTATION

During the GIS Implementation Stage, it is useful to spend some time defining the Key Performance Indicators (KPIs) against which the success (or otherwise) of the GIS will be measured periodically. These KPIs should be based around the benefits which were outlined in the Cost/Benefit Analysis and Business Case so that senior management:

- can see that there are 'goalposts', that they are business-based and that they have the focus (attention) of the GIS group;
- are fully aware of when a 'goal has been kicked' – that is, that a business benefit has been attained; and
- can use these KPIs as the basis for reporting progress.

The KPIs should be communicated to senior management and to all staff, and be in a form that is clear and easily understood with timeframes (where appropriate), i.e.:

- we will complete the data conversion by 1 July;
- all of the planners will have full GIS training by 1 August;
- all planners will be using the GIS for planning decisions by 15 September;
- planners should be making decisions 5% faster by 1 December;
- all assets data will be made available to the GIS team by 1 November;
- all assets staff will use the GIS for planning tasks by next February;
- the GIS will provide information for asset inspections by March next year.

Note that the these KPIs should also clearly enunciate if there is an expectation from other stakeholders that there are deliverables expected from them (i.e. 'assets staff will provide assets data'). This is particularly important so that if a KPI is not met (because of non-delivery from others) the senior management are aware of the reason why that specific KPI was not met.

Therefore, these 'goalposts for the business' provide the basis for reporting how the GIS (and the GIS group) is performing with respect to its KPIs. If performance has been good (as should be the case if all proper planning processes have been followed) then this will be a major factor in gaining additional budget or approvals for new initiatives. Conversely, if the KPIs are not met, the question 'why not?' often follows. If the latter is the case, having documented 'evidence' as to why this has occurred (e.g. 'assets staff did not provide assets data') could be important to long-term involvement with the project.

11.4 IMPLEMENTING AND 'SETTING TO WORK' OF THE GIS

At a suitable time during the process of physical implementation of the hardware and software, training of all personnel and commencing the data capture and conversion, the GIS implementation should start to 'process work' and to complete tasks for the organisation.

This is usually a 'settling-in time' when it is important to ensure that staff are focused on their tasks and that the expectations of other stakeholders in the organisation are continually reinforced at the correct level. There is a risk that when GIS is implemented some stakeholders will view the GIS as the panacea that is going to correct a number of problems inherent with their work processes or data – particularly without them having to do any work. As such, there needs to be continual reinforcement of the inputs, outputs, expectations and effort required in order to ensure that all stakeholders in this process are 'on the same bus'.

If there are no major technical or organisational problems and if KPIs are being met, then this stage is close to the end of the process, with only periodic reviews needing to be undertaken to continually ensure that everything is 'on track'.

11.5   UNDERTAKING A POST IMPLEMENTATION REVIEW

At a suitable time after implementation of the GIS (e.g. 12 months) it is useful to have a Post Implementation Review (PIR) undertaken by an independent person to ensure that all of the deliverables from all contracts have been delivered and that the GIS is meeting the business aims as envisaged in the original Strategy and Cost/Benefit Analysis.

This usually involves correlation of the KPIs to the original Strategy and Cost/Benefit Analysis, review of the content and functionality of the GIS and the database used and review of the timeliness and efficiency of staff to undertake the tasks required. In all, it would be expected that a PIR would highlight issues and tasks which need further attention, the risks of not addressing those issues and the possible mediation that could be implemented to minimise those risks.

The PIR should also be able to be used to reinforce to management that the investment that the organisation has made in GIS is being taken very seriously, that the staff are focused on delivering business outcomes (through KPIs) and that there is a good correlation between the contracts (as entered into) and what was delivered (software, services, data, etc.). The PIR should also be about celebrating successes! This is, of course, based on the assumption that there is a success to celebrate.

From time to time (e.g. at 3 or 5 years) it may also be useful to undertake an overall review of the GIS performance compared to business requirements, noting that the software market would have 'moved on' by that time, as would the organisation. The requirements of a more mature GIS user organisation invariably change, and expectations from their GIS invariably increase. The correlation of 'organisational expectations' of the GIS against the capability of the GIS to deliver those expectations should be continually re-appraised.

11.6   BENCHMARKING

Most organisations (particularly government and local government) have continuing pressures to improve performance and become more efficient and more customer-focused, i.e. to provide 'more for less'. One of the most effective ways to focus on business improvement is to aim for 'Best Practice' and compare performance against that of other 'like' internal or external groups.

However, while internal review can be useful for efficiency, it is of little use to an organisation to know that 'you are doing the wrong things really well'. Therefore, external comparisons have often been found to be a more effective benchmarking tool than internal comparisons.

In order for comparisons to be effective, there obviously needs to be some form of measurement (i.e. metrics) which can be applied to determine a measure of 'better performance'. Benchmarking across several organisations is often a useful method to measure performance for further correlation. Some (non-GIS) categories which are typically benchmarked are:

- the cost to issue a rates notice;
- the cost of child care services;
- the response time for processing applications;
- the length of customer service queues;
- the response time to repair services.

Therefore while comparisons of this type are useful to create 'Common Practice', it is 'the best of the common practice levels' which is often referred to as 'Best Practice'.

The surveys presented in this book are therefore based around these concepts – to measure industry trends so that, over time, common practice can be derived and form the basis for developing a measure of best practice.

Therefore, undertaking benchmarking studies which measure a number of performance indicators (metrics) for a small number of business processes used by a group of 'like' organisations (e.g. 6–10 water utilities or local governments) has the potential to produce quality indicators of common and best practice which participants can then use to determine their overall performance (i.e. are they a 'leader' or 'lagger' in each category).

Thus, one of the most successful methods of 'getting it right' is to measure the organisation's performance with that of a group of other 'like organisations' and to develop a 'road map' of where the organisation is and should be going. This will ensure that the right processes are used to provide the basis for significant improvement and a reduction in costs.

11.7   SUMMARY

In summary, the 'Implementing GIS' stage should be a culmination of the earlier specification and Tender processes and be the 'end of the start' and not the 'start of the end'. The KPIs and PIR should provide the basis for reporting and re-evaluating progress against that which was originally envisaged with a view to ensuring that the GIS continues to meet the requirements of the business well into the future.

Section 2.3 outlined that there is a high amount of product 'churn' in the market with a consequent lack of product loyalty, due to a number of factors, the principal ones being:

- the GIS not being able to meet the changing business needs of the organisation;
- the GIS not meeting the needs of an organisation which is split/merged with a different organisation with different GIS technology/processes/business needs;
- the GIS being too expensive or providing poor support; and
- the GIS system withdrawing from the market (e.g. three major GIS vendors have withdrawn from the Australian/New Zealand market in the last 5 years).

This churn will invariably result in the need to renew, upgrade and retrain staff and organisations in the use of SI systems to meet specific business needs. As such (and like the IT industry in general), this is a cycle which is destined to continually repeat, refine and replace – and adequate procedures need to be put in place to make sure that this is an ordered process for customer organisations.

12   The Best and the Worst

No book on GIS would be complete without a quick look at the best and the worst of GIS products and projects. The following examples are just 'several of many', but ones which stand out – coincidentally the Best example of a GIS product has been undertaken by a private company with obviously extensive research and the Worst example has been undertaken by a Government agency with obviously little research into the needs of the target market.

12.1   AND THE BEST IS . . . GOOGLE EARTH

The best GIS product to be released in a long time would have to be Google Earth – never has there been a product in this industry that is so intuitive and easy to use, with the potential to re-define the way people look at geography.

This is because:

1. It is so intuitive and easy to use (have I said this already?).
2. It combines an enormous number of images to present 'the earth' as a globe and in such a manner that it would be able to be understood by everyone, even those who have no knowledge of geography and maps.
3. It has 'spawned' a sub-culture developing Google-maps on all sorts of subjects:

   a. http://beermapping.com/us-brewery-map/ – as the name implies, always very handy;
   b. http://www.mackers.com/projects/dartmaps/ – Ireland trains in action;
   c. http://grad.icmc.usp.br/~cipriani/bighole.php – where you come out when you dig a hole through the earth – particularly useful to know after one has had a few beers;
   d. http://bbs.keyhole.com/ubb/showflat.php/Cat/0/Number/175202 – the Google Earth collection of crop circles, coincidentally proving that not all crop circles are formed by aliens.

While these examples may seem trivial on first inspection, think closely about what is being demonstrated here.

4. It exposes a large number of people to GIS who may never have been exposed to spatial technologies before – generally this will occur when father or mother watches son or daughter using Google Earth and thinks 'I could use this at work, if it had our data . . .'.
5. It provides a (common) forum to add spatial data from Government and Local Government GIS to turn Google Earth into an 'on-line GIS'.

Achieving Business Success with GIS   Bruce Douglas
© 2008 John Wiley & Sons, Ltd   ISBN: 978-0-470-72724-9

While the future development path of Google Earth is unknown, it is likely that there will be more data at better scales with more functionality. This has the potential to revolutionise the GIS industry and perhaps become the 'killer application' of this decade. The reaction from Microsoft, Yahoo, etc. has been tentative, but I am sure they will provide the basis for further extensive development for this genre of product.

12.2   AND THE BEST IS (ALSO) . . . WEB 2.0

And the Google Earth best product would have to lead (or some might say follow) recent disruptive industry developments generally referred to as 'Web 2.0' (see O'Reilly).[1]

While the ramifications of the Web 2.0 phenomenon are not yet fully understood, early signs are that it is very significant and could lead to harnessing 'Collective Intelligence', the central principle behind the success of the early Internet companies who have survived to lead the Web 2.0 era.

This collective intelligence includes hyperlinking as the foundation of the web, portals to form collective works, continuing developments in search engines such as Google's PageRank, eBay's organic growth in response to user activity and enabler of a context in which that user activity can happen, the continued success of sites such as Amazon who have made a science of 'user engagement' and the phenomenon which is Wikipedia, an online encyclopaedia based on the unlikely notion that an entry can be added by any web user.

As such:

> Web 2.0 is a phenomenon where the threshold of technology is vastly lowered, resulting in greater public participation in the use, sharing and creation of information, including spatial data. This is also impacting on traditional GIS vendors in a very significant manner as they are suddenly faced with tremendous competition from Web 2.0 and open source developments which may not only push them back firmly into their traditional professional market, but also takes away a slice from their 'home turf' by consumer market forces. Overall this isn't really that noticeable yet as the total spatial market (professional and consumer) is still rapidly expanding, but it is significant nevertheless. Web 2.0 is creating a whole new consumer market which is being conquered by new players – not just Google, Yahoo or Microsoft but also private citizens creating and sharing their own data and systems, all free at the point of use.[2]

O'Reilly goes on to explore the seven principles of Web 2.0 with demonstrations of key principles summarised as 'core competencies' for Web 2.0 companies:

1. Services, not packaged software, with cost-effective scalability.
2. Control over unique, hard-to-recreate data sources that get richer as more people use them.
3. Trusting users as co-developers.
4. Harnessing collective intelligence.
5. Leveraging the long tail through customer self-service.
6. Software above the level of a single device.
7. Lightweight user interfaces, development models *and* business models.

---

[1] http://www.oreillynet.com/pub/a/oreilly/tim/news/2005/09/30/what-is-web-20.html.
[2] Thierry Gregorius, Head Data Management & Geomatics, Global Exploration, Shell International Exploration and Production, The Netherlands.

The web as the platform has only recently become a credible notion, and in reflection of where the IT and GIS industry has come over the last 20 years it is one which is going to continually re-invent the science of spatial information, particularly with companies such as Google leading the pack.

12.3 AND THE WORST IS . . .

Undoubtedly the worst GIS product that I have seen for a long time would have to be the Australian Federal Government 'National Toilet Map' at http://www.toiletmap.gov.au/, produced by the Department of Health and Ageing.

While a National Toilet Map is an excellent concept and is something that has been needed for a long time, its target audience is people who suffer from incontinence, with the web site stating that *'incontinence is a common health problem that affects over 2 million Australians of all ages and backgrounds.'* However, I would expect that a high percentage of these 2 million incontinent people would also be part of the estimated 5 million elderly people who often need toilets when travelling. A Toilet Map would be very useful for these people.

Unfortunately, while the concept is good, the delivery of the information leaves a lot to be desired. In my experience (certainly from discussion with elderly relatives), most elderly people do not have computers and are not Internet literate, although a few are starting to become so. And of those that do use computers, I strongly doubt that they would be using them while motoring down the highway. Therefore, the time that one most urgently needs directions to a toilet (i.e. being 'caught out' when driving) is also the time when the required map is least able to be accessed using this service.

And even if one were foresighted enough to print out the 300 map pages for the drive from Sydney to Melbourne, and had a backseat big enough to accommodate the bookcase required, the person would still need to use a street directory to find the actual toilet – i.e. one would need another (more useful) map so that the first map could be used.

There are a number of problems with this product, the major ones being:

- using the 'Trip Planner' option, the toilets found for the route are shown using 'given directions', i.e. 'go 270 metres and turn right in Smith Street' etc. – this works well if the directions are followed explicitly and if a detour is taken to visit *every* toilet on the route (i.e. all 485 toilets between Sydney and Melbourne, several in each small town). However, the trip planner is less useful if one chooses to not visit *every* single toilet but misses (say) 450 when the urge is not apparent, or if a detour is taken to do some sightseeing – in which event a street directory would be required to get back to the toilet route;
- using the 'Find Toilets' option, the toilets for each locale (usually a town or locality) are listed separately with a street address, and are shown on a map at a scale that makes it unreadable, with no useful street names shown and with symbols overlapping most of the street network, therefore requiring one to use a street directory to actually find the toilet.

This is also an example of where a high web site 'hit-rate' would not be indicative of success. I would expect that a high hit-rate (should there be one) would represent users recording a hit every time they tried to print out the 485 toilet maps for the drive (just in case).

So, in summary, while the Toilet Map is an excellent idea it has unfortunately been let down by its lack of understanding of the needs of the target audience, i.e.:

- the use of technology which is inappropriate for the target audience; and
- showing data at too small a scale and with a lack of nomenclature such that another (more useful) map is required so that the first map (the Toilet Map) can be used.

I would respectfully suggest that something like a series of paper 'strip maps' themed on toilets would have been more appropriate, something like those available from motoring clubs (and much loved by motorists) and distributed to incontinent and/or elderly people in booklets (perhaps regionally based). I would expect that this 'low-tech' paper format would have a high take-up rate without the (toilet) user having to purchase a computer and printer, learn how to use the technology, install an Internet connection and do a few hours of research just in case he/she is 'caught out' on the Sunday drive.

Conversely, if the Government provided the toilet locations free of charge to the street-mapping vendors, they could be pre-loaded into satellite navigation systems, again assuming that elderly drivers used these devices (which I suspect they do not).

Another example of our tax dollars at work . . .

13 Closing Remarks

The implementation of a high-technology solution such as GIS in an organisation can be very successful and can significantly enhance the business of the organisation implementing it (the GIS) if it is approached in a sensible and methodical manner, following a number of tried and proven processes. However, the implementation of GIS can be fraught with difficulties, as outlined in this book.

However, although there are difficulties, the secret of successfully implementing GIS is to:

- develop the business and technical requirements for the GIS carefully and document them in a specification which is agreed to and signed off by all stakeholders;
- develop a clear and unambiguous business case (i.e. financial need) to implement GIS to deliver specific and quantifiable business benefits and convey this to senior management in an open and concise manner;
- undertake a clinical evaluation of the suppliers in the market-place using a fair and equitable process to select a system which most closely aligns with the business requirements and does so in a cost-effective manner;
- implement the GIS to focus on the business requirements which are backed up by documented 'goalposts' based on a comprehensive implementation plan; and
- undertake a programme of continuous improvement to continually re-focus the GIS onto (continually) changing business needs without being encumbered by the need to blindly follow a specific product loyalty path.

If the above process is followed in a careful manner, with due attention paid to detail and probity, the resulting outcome should be the acquisition and use of GIS technology which will provide significant business benefits to the host organisation.

If the above processes are not followed or are 'glossed over', the GIS will almost certainly not meet the business needs and will not be considered successful, and when that happens senior management will probably search for the person who initiated the process. It is therefore in the interests of the person initiating or managing this process to ensure that it is done correctly and diligently.

Industry experience and research suggests that most GIS implementations fail because of organisational issues, management issues or political issues. Note that technology issues are not mentioned in this list – any technology will almost certainly be successful if the organisational/management/political framework is correct. Conversely, any technology will almost certainly fail if these issues are not addressed correctly.

In closing, I trust that this book has been of some interest in planning and implementing spatial information systems. There is no 'right or wrong' answer for making GIS successful – just common sense, good management, good organisational structure and (a fair amount of) experience.

Achieving Business Success with GIS   Bruce Douglas
© 2008 John Wiley & Sons, Ltd   ISBN: 978-0-470-72724-9

Glossary

ANZLIC	Australian and New Zealand Land Information Council
ASDD	Australian Spatial Data Directory
CAD	Computer-Aided Drafting (Design)
COE	Common Operating Environment
CSF	Critical Success Factor
DWG	Drawing file format of AutoCAD
DXF	Direct eXchange Format (file format of AutoCAD)
ECW	Enhanced Compression Wavelet (format for Imagery)
EOI	Expression Of Interest
FME	Feature Manipulation Engine (for spatial data translation)
FRS	Functional Requirements Specification
GDA	Geographic Datum Australia
GIS	Geographic Information System
GITA	Geospatial Information and Technology Association
GML	Geography Markup Language
GPS	Global Positioning System
IE	Internet Explorer (Microsoft)
IGES	International Graphics Exchange Standard
IT	Information Technology
KPI	Key Performance Indicator
KRA	Key Result Area
LAN	Local Area Network
LBS	Location-Based Services
NPV	Net Present Value
OGC	Open Geospatial Consortium
PDA	Personal Data Assistant
PDF	Portable Document Format (file format of Adobe Acrobat)
PIR	Post Implementation Review
RDBMS	Relational DataBase Management System
RFT	Request For Tender
ROI	Return On Investment
SDTS	Spatial Data Transfer Standard
SI	Spatial Information
SOE	Standard Operating Environment
SWOT	Strengths, Weaknesses, Opportunities, Threats
USGS	United States Geological Survey
WAN	Wide Area Network
WFS	Web Feature Service
WMS	Web Mapping Service
XML	eXtensible Markup Language

Achieving Business Success with GIS   Bruce Douglas
© 2008 John Wiley & Sons, Ltd   ISBN: 978-0-470-72724-9

Index

Achieving Business Success with GIS   Bruce Douglas
® 2008 John Wiley & Sons, Ltd   ISBN: 978-0-470-72724-9